In 2002, John Robertson became a volunteer for the Alnwick Garden in Northumberland. In late 2003, he was asked to undertake research into the stories of the plants planned for the Poison Garden, then under construction. In 2005, when the garden opened, he was asked by the Duchess of Northumberland to become Poison Garden Warden. John trained staff and volunteers to take Poison Garden tours, as well as taking tours himself and giving talks in the conference suite at the garden. For his own interest, he continued to research the facts, myths and stories about poison plants. He currently runs the web site www.thepoisongarden.co.uk and continues to give talks. He is a member of the Society for the Study of Addiction.

IS THAT CAT DEAD?

And Other Questions About Poison Plants

John Robertson

Book Guild Publishing
Sussex, England

First published in Great Britain in 2010 by
The Book Guild
Pavilion View
19 New Road
Brighton, BN1 1UF

Copyright © John Robertson 2010

The right of John Robertson to be identified as the author of
this work has been asserted by him in accordance with the
Copyright, Designs and Patents Act 1988.

All rights reserved. No part of this publication may be reproduced,
transmitted, or stored in a retrieval system, in any form or by any
means, without permission in writing from the publisher, nor be
otherwise circulated in any form of binding or cover other than that
in which it is published and without a similar condition being imposed
on the subsequent purchaser.

Typesetting in Garamond by
Keyboard Services, Luton, Bedfordshire

Printed in Great Britain by
CPI Antony Rowe

A catalogue record for this book is available from
The British Library

ISBN 978 1 84624 425 4

Contents

1. What's it All About? — 1
2. Why Are Plants Poisonous? — 13
3. What's the Most Poisonous? — 27
4. What is the Biggest Killer? Part 1 — 39
5. What is the Biggest Killer? Part 2 — 51
6. Have You Got Something Undetectable? — 65
7. But Aren't These Used as Medicines? — 77
8. Have They Been Used as Murder Weapons? — 91
9. Do You Mind Me Asking? — 107
10. Does Mandrake Really Scream? — 117
11. Why Don't We Buy the Poppy Crop in Afghanistan? — 129
12. Should Local Councils Grow the Castor Oil Plant, Ricinus communis? — 137
13. Where's the Cannabis? — 149
14. Why Are Some Plants in Cages? — 171
15. Why No Fungi? — 179
16. Is That Cat Dead? — 189

17	What is Rosemary Doing in a Poison Garden?	203
References		213

1

What's it All About?

'... so, remember, while you're in the Poison Garden, don't touch, don't pick, don't eat and don't smell. There are plants in this garden that will make you ill just from the smell.'

At which, a tall teenage boy, towering above the rest of the otherwise unremarkable group of 20 or so visitors to the Alnwick Garden, in Northumberland, took a step away from his mother and said, 'That sounds like my brother'.

While I was working at the Alnwick Garden, a number of my colleagues suggested I should write a book based on all the research I had done for four and a half years. Once I was no longer working in the garden, I had the time to think about whether such a book might be of any interest and, finally, it occurred to me that by trying to answer the questions most often asked by visitors to the Alnwick Garden Poison Garden I might be answering the questions most people have about poison plants – and that might make the book more enjoyable and more pragmatic than a simple A to Z of theoretical capabilities.

And so this book will deal with the questions that have been asked by many of the hundreds of thousands of people who've passed through the garden. Undoubtedly, one of the most interesting aspects of being Poison Garden Warden at the Alnwick Garden was listening to visitors, either asking questions or sharing their experiences of poison plants (or smelly brothers).

With hindsight, I regret that there never was a complete count of the number of visitors coming to the Poison Garden as it would be interesting to know the exact number and have a breakdown of who they were and how many came more than once. In the absence of a precise tally, the best I can do is estimate that, of the nearly 500,000 visitors who are thought to have come to the Poison Garden from the time it opened to the time I left, around 100,000 listened to one of the nearly 5,000 tours I took around the garden.

With that many people, it is inevitable that some questions would crop up again and again; that there would be, in the requisite jargon, 'FAQs' – Frequently Asked Questions.

My purpose in writing this book is to answer some of the most frequent of the FAQs and, in so doing, look at the myths, folklore and facts about poison plants.

Poison is fascinating. Who doesn't find the thought that the loaf bought fresh from the baker might contain substances capable of causing hallucinations and damaging the circulation so as to cause gangrene slightly disturbing? Who hasn't looked at an unknown autumn plant laden with bright attractive berries and wondered if it would be safe to pick and eat them? Crime writers who've had their villains employ poison in devious and innovative ways and made their fortunes as a result know that people like to hear about poisons and the people who use poisons.

But poison is also one of those things you don't like to admit an interest in. Perhaps that's why the first question many people asked was, 'What's the Poison Garden about?' as if drawing attention to poisons was something a bit suspect.

The answer I always gave to this question was, 'It's all about stinging nettles'.

A sting in the tale

Urtica dioica is known as 'common nettle', 'stinging nettle' or 'the Naughty Man's plaything' (the Naughty Man being the devil).

WHAT'S IT ALL ABOUT?

Everybody knows that brushing against a stinging nettle will cause the skin to become hot and painful and may cause red swellings to appear. Strangely, many people don't seem to see this as poisoning, perhaps because they are used to adding the epithet 'deadly' to the word poison and the effect of the nettle, though irritating and uncomfortable, is only in very extreme cases serious.

The sting is caused not by formic acid, as often believed, but by a combination of histamine, acetylcholine and 5-hydroxytryptamine. The plant carries tiny needles each with a sac of poison at its base. Brushing the plant initially breaks the end of the needle and then the bending action squeezes the sac and drives the poison into the skin. Anyone brushing against Urtica dioica is, in effect, giving themselves hundreds, possibly thousands, of injections of poison. Since it is the sideways brushing action that breaks open the sacs of poison, there is an element of truth in the old saying that one should 'grasp the nettle' as this is supposed to flatten the tiny hairs that carry the stinging compounds and, thus, prevent the sting.

The belief that deliberately inducing a nettle sting will reduce the pain of arthritis is widely held and has a long history. It may also have some basis in fact. In June 2000, researchers from the University of Plymouth published the results of a randomised controlled double blind study of the effect of nettle sting on arthritic pain at the base of the thumb.

The trial involved 27 people who were given a non-flowering plant, either Urtica dioica or Lamium orvala, deadnettle, and asked to apply the underside of one leaf to the skin on the base of the thumb or other painful area once a day for thirty seconds. The patients were warned that they might experience stinging but were told the plants in the trial were all stinging nettles.

All the patients using stinging nettles reported an almost immediate reduction in pain and their diaries revealed a significant reduction in analgesic medication. As with many such trials, the researchers concluded by calling for further work on a larger scale but, to date, nothing more has been published.

This belief in its pain relieving, and prophylactic, effects leads to a widespread, but almost certainly untrue, piece of folklore. Many sources say the stinging nettle was introduced to Britain by Roman soldiers who expected to need to beat themselves with nettles to create a burning on the skin that would help them keep warm in the British winter. Given that the nettle dies back in the winter, it would not have been available when it was most needed, so either the story is untrue or the Romans were ignorant of the conditions they would meet.

In his *The Englishman's Flora* (1958), Geoffrey Grigson talks about using nettle fibres to make cloth. He cites the 18th-century Scottish poet, Thomas Campbell, as writing of sleeping in nettle sheets and eating off a nettle tablecloth. Grigson also says that nettle cloth was found wrapped around cremated bones in a Danish grave from the later Bronze Age, suggesting the plant was known in northern Europe long before the Romans arrived.

Grigson (1958), John Gerard (in his 1598 *Great Herbal*) and others call Urtica pilulifera the 'Roman nettle'. Grigson says it was found in Romney in Kent where Julius Caesar landed but then points out that its discovery came long after it had been given the name Roman nettle and 'long after' it had been grown by Gerard. Grigson describes its alleged use for keeping warm by Roman soldiers as 'the tall story'.

He does, however, say that it seems to follow mankind around and talks of it appearing in New England in the 17th century after the arrival of British settlers. This 'following man around' leads to the belief in the Highlands and Islands of Scotland that nettles grow from the bodies of the dead. Even recently, it has been said that searchers in former Yugoslavia seeking hidden graves resulting from 'ethnic cleansing' would target areas where nettles grew taller than average as signs of human remains underneath.

Pliny the Elder, whose 37-volume *Natural History* was written in the first century and contains almost all the knowledge in the Roman world, much of it quite trivial, has a lot to say about the uses of the nettle. Yet, even though he served in the Roman army in Germany, he makes no mention of its use as a personal heating system.

John Parkinson, in his *Theatrum Botanicum* (1629), is believed to have been the first to claim that nettle was an import from Italy. William Camden in his 1806 book *Britannia* says that the Romans believed Britain to be a place where giants lived in the mountains and serpents roamed the valleys and a person would need to be mad to voluntarily travel to Britain. He then gives the story of the Romans bringing nettle seeds with them.

It may be, therefore, that the Romans brought nettle to Britain not knowing it was already here and not knowing it would die off in the winter. Equally, nettle may just have come with them in the way that it followed the early settlers to New England.

There is, of course, the added complication of how Roman were the 'Romans'? The Roman Empire depended on assimilating conquered peoples, so many soldiers of the invading army were native to other parts of the Empire. Furthermore, by the time Hadrian's Wall was built about 150 years after the initial invasion, many of the garrison soldiers were native to England and, thus, well used to the climatic conditions.

On balance, Grigson's view that soldiers beating themselves with nettles was a 'tall story' is, most probably, correct.

In addition to its alleged effect on arthritis and rheumatism, the stinging nettle has been credited with a variety of other curative properties. Generally it is believed to remove curses, allay fear and avert danger; and in Denmark, the stings are said to protect against sorcery. It is also said to reduce prostate enlargement and, thus, have aphrodisiac properties.

In *A Modern Herbal*, the 1931 guide to medicinal plants, it is said that 'as an arrester of bleeding, the nettle has few equals'. (This tome is attributed to Mrs Grieve but is actually the work of Maud Grieve and Hilda Leyel, founder of the Society of Herbalists, who added American herbs to broaden the book's marketability.) One wonders if its use in this way was based on the observed behaviour of animals. An elderly visitor to the Poison Garden talked of keeping pigs when he was in his teens. When he needed to slaughter a young pig for pork, the accepted method, at the time, was to cut the pig's throat and let it bleed to death. He said that whenever he did so the pig would rush to the nearest nettle patch and begin eating and rolling in it.

Hildegard of Bingen, the 12th-century mystic, composer and writer, suggests nettle as a cure for forgetfulness. The recipe says pound the leaves to a juice, mix with a little olive oil and apply to the chest and temples of a forgetful person during sleep. Do this often and the forgetfulness will gradually lessen.

So, Urtica dioica is a complex and fascinating plant but why do I say it is at the core of the Poison Garden? Because almost everybody knows that if you get stung by a nettle you should immediately rub the area with a dock leaf. Almost everybody knows this because it gets passed from generation to generation but it is, these days, virtually the only piece of plant folklore that is handed down.

The Poison Garden is about those stories that are no longer passed on; the things parents would tell their children as bedtime stories to send them to sleep with their heads full of fairies and magic but also, and more importantly, the tales they would tell to keep the children away from a harmful plant.

'Harmful' is a very significant word when it comes to poison plants. A plant can be as poisonous as it likes but, if there is nothing about it that encourages you to engage with the poison, it can never be harmful.

The dockleaf conundrum

Let's return to Rumex obtusifolius, the dock, broad-leaved dock or, in Scotland, docken. As well as being the alleged cure for the sting of the nettle, of which more shortly, Rumex obtusifolius is a poisonous plant. The toxic components are organic oxalates in the form of needle-shaped crystals, which can irritate the skin, mouth, tongue and throat, resulting in swelling, breathing difficulties, burning pain and stomach upset. The swelling of the mouth and throat could be severe enough to close the airways and prevent breathing.

Once ingested these organic oxalates preferentially bind to calcium in the body. Regular ingestion of small amounts could lead to calcium deficiency and to the build up of kidney stones if the calcium oxalate formed is not excreted.

But, there is no record of anyone ever suffering calcium deficiency as the result of eating dock and, certainly, no record of anyone dying of asphyxia after eating a large amount. Our knowledge of the effect of oxalates comes, mostly, from their presence in other plants though there have been instances of animal poisoning resulting from the ingestion of large amounts of other species of Rumex. What stops people being poisoned by eating dock is that there is nothing about the plant that would encourage you to eat it. The leaves look tough, and are, and the taste, even of a young leaf, is sour. The flowers and the seeds on the mature plant are dingy and don't recommend themselves.

In spite of this limited harm to animals and absence of harm to humans, Rumex obtusifolius was one of the five plants named in the Weeds Act of 1959. The other four are Senecio jacobaea, common ragwort, Cirsium vulgare, spear thistle, Cirsium arvense, creeping or field thistle and Rumex crispus, curled dock. As far as Senecio jacobaea, ragwort, is concerned, the requirements of the Weeds Act, 1959 have been superseded by the Ragwort Control Act, 2003, of which more in Chapter 2, but, for the other four plants, its provisions, theoretically, still apply. It is important to note that the Act does not make it illegal to grow these plants but gives the Secretary of State for Agriculture and Rural Affairs the power to issue an order for their control if it is perceived that there is a danger of them spreading, especially onto land used for grazing animals or for producing animal feed.

IS THAT CAT DEAD?

The intention of the Act was to protect agricultural production but, by 1984, the government of the day acknowledged that developments in the use of herbicides and changes in farming practice had all but rendered the Act redundant. Its powers had only been invoked twice in the previous five years. On balance, however, according to a parliamentary answer recorded by Hansard (the edited daily record of what is said in both Houses), it had been decided to keep the Act on the statute books as it proved a useful incentive to persuade land occupiers to deal with these plants when they became problematic.

So, Rumex obtusifolius remains a theoretical outlaw but many people who use it to treat nettle stings would not see it that way.

As well as the main FAQs, which merit a chapter of their own, there are many other questions that have been asked about the plants in the Poison Garden. Two of these were 'Does dock really work?' and 'How does dock work?' The answer to the first question is 'It seems to for many people', though not all, because a number of visitors assured me that they had used dock on nettle stings without achieving any relief. As to the second question, there does not seem to be an answer. Various mechanisms have been suggested, like the slightly alkaline juice of the dock neutralising the slightly acid stinging chemicals, or the coolness of the leaf bringing relief. I've even read the suggestion that vigorous rubbing of the area overloads the nerve channels so the brain has to pay less attention to the older pain signals but I have not been able to find any formal research into the matter. Incidentally, all the proposed mechanisms would seem to work with just about any leaf and do not explain why the dock became the leaf of choice.

It is often said that you will always find dock growing close to stinging nettles as if that is some part of nature's grand design, but such a belief is simply an example of the arrogance of the human race, which we will look at in more detail in later chapters. The reality is that Rumex obtusifolius is a very common plant, as is Urtica dioica, and any piece of uncultivated ground is almost certain to have both of them present.

WHAT'S IT ALL ABOUT?

Not that rubbing a dock leaf is the only 'cure' for nettle sting. Ray Mears, the TV presenter who is an authority on the subject of bushcraft and survival, maintains that rubbing the top side of a nettle leaf, the side with no stinging hairs, onto a sting will end the pain but handling a nettle leaf to apply it in this way runs the risk of creating new stings on the fingers. A visitor to the Poison Garden told me her local herbalist said that eating raw nettle leaves immediately after being stung would stop the pain.

I've experimented with this and found that it does appear to work. I've tried eating some leaves immediately after brushing against the plant and eating leaves up to fifteen minutes before stinging myself. My final experiment was to sting myself and then try eating leaves fifteen minutes after to see if the sting would subside. The only trouble was that, in spite of rubbing myself firmly enough to raise red lumps on the skin of my thigh, I felt no sting. This brings up the subject of the placebo effect where the brain does the curing by itself, something which I'll discuss in more detail in the Chapter 7 when we look at medicine, herbal medicine and homeopathy.

It's all about drugs

Whilst the main purpose of the Poison Garden was to retell the old plant folklore about poison plants and see how it came about and whether the plants really deserved their reputations, there was also the question of substance abuse. The more I researched the plants in the garden and the more stories I heard from visitors about their experiences, the more I came to see that the Poison Garden was all about drugs. I would say it is possible to make a substance abuse point about any of the plants. Take, for example, the stinging nettle, Urtica dioica. If everyone knows that brushing against the plant will lead to burning pain, why do people still get stung by it? What is there in the human character that means we are willing to take risks with harming ourselves even if we are well aware of the harm that can arise? Do we think we are different? 'If I'm careful I can walk through these nettles either side of the path and not get stung.' Do we want the result regardless of the suffering? 'I have to walk along this path because I must get to my destination.' Do we deny the potential for harm? 'Those are deadnettles, not stinging nettles.'

What a lot of rhubarb

Take, as another example, rhubarb. There are so many edible species and varieties of the genus Rheum it is easier to use the common name, rhubarb. Though the stalks are widely eaten as a dessert, the leaves are highly toxic. They contain oxalic acid in the form of oxalates, which is widely believed to be the poison, but there is evidence that these are insufficient to cause the known fatalities. There are reasons to believe that anthraquinone alkaloids are to blame, but the mechanisms have not been fully studied. That makes it a useful example of how little we know about how these plants and the substances they contain work, which is an important consideration when discussing the effects of cannabis, as we will in Chapter 13.

Symptoms of rhubarb poisoning are said to include weakness, burning in the mouth and throat, breathing difficulty, pain in the eyes and stomach, nausea, vomiting, diarrhoea, seizures, red coloured urine and kidney stones. Death from eating rhubarb leaves occurs quickly and is preceded by drowsiness, possibly leading to coma, convulsions, internal bleeding and nosebleeds as coagulation is inhibited. Symptoms begin within an hour of ingestion.

Mrs Grieve talks about a letter in the *Gardener's Chronicle* for 1846 in which the gardener of the Earl of Shrewsbury at Alton Towers, Staffordshire, told how rhubarb leaves had been used there for many years as a vegetable. Then he wrote again to correct it and say it was the leaf-stalks he meant. It is possible that this error might have been passed down, leading to the misuse of the leaves.

There are a number of references to problems during World War I when rhubarb leaves had been recommended as a vegetable until a number of deaths occurred and the advice was withdrawn. In spite of this, it is said that there were twelve deaths in France during World War II as a result of eating rhubarb leaves, and on the BBC's 'People's War' website a woman writes about being 12 years old in occupied France and eating rhubarb leaves as a vegetable until there was an announcement in the press that this could cause poisoning.

Though stories told by visitors to the Alnwick Poison Garden were always interesting to hear, they were frequently wrong. Rhubarb provides two good examples of such stories.

A visitor claimed that Captain Cook had been the first person to realise the role of green vegetables in preventing scurvy, which meant he had the first scurvy-free vessel in the navy. Unfortunately, when other green leaves were not available he fed his crew on rhubarb, leading to many deaths. In fact, Cook's vessel suffered almost as many scurvy deaths as other vessels in the fleet in spite of his attempts to provide fresh fruit and vegetables. I can find no reference to his men being fed rhubarb leaves.

Another visitor stated, as fact, that the mixture of rhubarb and pineapple is deadly. This is not so. There are numerous recipes available for rhubarb/pineapple pie. The suggestion is that the acid in pineapple combines with the small amount of oxalic acid in rhubarb stems to produce a harmful strength. The acids in pineapples, which are not oxalic acid, could break down the cell walls of rhubarb and increase the bioavailability of the small amount of oxalic acid in the rhubarb. This could increase the acid taste but would not affect the absorption.

And the substance abuse message from rhubarb is simply this: that used one way rhubarb is pleasant and enjoyable, but used wrongly it can kill, something that can just as easily be said for alcohol.

So every plant in the Poison Garden has something to tell us about our use and abuse of substances.

And, of course, the Poison Garden also has quite a lot to do with sex. Because it's all about our relations with the plants and each other, sex is an essential part of the story. Naturally, when talking to the normal groups of visitors, often including quite young children, there was not a lot of opportunity to talk about the sex but it will get more than a passing mention as we move on to the next and subsequent chapters.

2

Why Are Plants Poisonous?

It was the week before Christmas and Father Christmas was in his grotto, otherwise known as the Treehouse pod. A family had come to the Alnwick Garden to take their daughter to see Santa but, before that, mum, dad, grandad and grandma wanted to tour the Poison Garden and that meant the little girl coming too. I'm not good at judging ages but I'd guess that she was in the range four to six years old, certainly no older than that.

To my surprise she paid quiet attention as we started our tour but, after about ten minutes, I noticed that her happy smile had gone and I thought she was getting bored. I asked her if she was all right and she looked straight at me, took a deep breath, and asked, 'Why do pretty flowers make people die?'

That question, more usually expressed in the form 'Why are plants poisonous?' continues to fascinate the human race. In fact, the answer has been known since the 16th century but it is so mundane that many people do not accept it and continue to look for some deeper reason.

The most usual answer given by people is that there is some evolutionary benefit to the plant from being poisonous often associated with furthering the line of the plant concerned. Take something like the cyanolipids contained in many fruit seeds, which, if macerated, can give rise to hydrocyanic acid, known in the past as prussic acid.

The trees in the Poison Garden were, primarily, there to give some shape to the garden by their height but, with some exceptions, it was possible to make the case that they contained poison. In one corner of the Poison Garden was a Malus 'John Downie', a crab apple. All apple seeds, more usually called pips, contain cyanolipids, which theoretically could give rise to cyanide, and most people who eat apples discard the core containing the seeds. Many of us were brought up believing that eating the core would cause a stomach ache and some young children were assured that eating the seeds would result in an apple tree growing in the belly.

There was another reason for having the Malus in the garden but it involved a bit of theological sidestepping. Though the Old Testament doesn't actually say what the tree in the Garden of Eden was, it is often assumed to have been an apricot. This is based on the Garden of Eden being somewhere in the Middle East and the apricot being a common tree in that area, the point of the fable being that temptation was all around and easy to succumb to. When the Old Testament was translated for European use, if the tree had remained an apricot then temptation would have been so exotic and unusual that no one would have been worried that they might be exposed to it. And so it became an apple tree but, because the modern cultivated apple is only a few hundred years old, it became a crab apple. If you accept those theological stretches, then the Malus 'John Downie' was the tree that brought about the downfall of the human race and, if that doesn't earn it a place in a Poison Garden, I don't know what would.

In fact, no harm will come from eating the pips of a single apple for the very good reason that the outer casing of the seed is very hard and passes through the system without being digested. There is one alleged case of poisoning by apple pips where the victim is said to have either deliberately crushed a bowl full of apple seeds to break down the outer skins or the time it took to collect a large quantity of pips meant that many of them had rotted, exposing the internal material.

The evolutionists' argument is that the tree ensures the survival of its line by producing a tasty fruit to encourage ingestion, which means that the excreted seeds will be distributed widely. The argument goes on to say that the seeds contain poison to discourage the creature that consumes the fruit from chewing them so that they cannot germinate when distributed. For this argument to work it is necessary to believe that if this bird, here, consumes a fruit, chews into the seed and dies of poisoning then that bird, there, and all its descendants, will know not to chew the seeds in the future. In fact, the evolutionary success is simply that the seed casings are not digested so the intact seed is excreted in a place where it has a chance of successful germination. If being poisonous has any evolutionary role it is an historic one rather than something that still assists the plants. Over the long period of evolutionary history, creatures with the ability to bite into the hard seeds of a plant like the apple tree have died out. It's also worth mentioning that plants whose seeds had such soft shells that they were digested rather than being excreted intact also died out.

Or the theory is put forward that a plant's foliage is poisonous to prevent it being eaten before the plant has a chance to complete its life cycle. That might be a reasonable proposition, though, just as with the seeds of the apple, it relies on some form of racial memory so that today's rabbit knows that, many generations before, one of its ancestors ate some foliage and died. Where it breaks down, additionally, is in the cases where a plant is still poisonous even when it is dead.

Ragwort – the misunderstood

Senecio jacobaea, ragwort, is best known as a poison for horses. The reason for this is simply that dead ragwort is as poisonous as living ragwort and that living ragwort has a feature that discourages consumption. Quite simply, living ragwort is extremely unpleasant to the taste and animals will ignore it if there is something else available. When I tried it, I found I couldn't finish chewing one small leaf because of its extreme sourness. As long as there is other, more palatable, grazing available, animals can be left in a field containing ragwort without concern.

That said, it has been found that, although the taste is so unpleasant, cattle will eat ragwort if there is nothing else available. Interestingly, when returned to normal grazing land these animals are found to continue eating ragwort. This leads to speculation that ragwort is addictive and, it should be noted, there are approximately 60 species of Senecio found in Mexico, many of which have been used by Mexican Indians for their psychoactive properties.

But, in general, ragwort only becomes a problem if it gets into hay as it has lost the taste but retains its toxicity after death.

The toxic components of ragwort are pyrrolizidine alkaloids, which are damaging to the liver. Liver damage is cumulative and symptomless until it becomes acute so regular ingestion of contaminated hay can lead to apparently sudden liver failure. The symptoms it causes are described by the names given to its effects: 'Walking Disease' and 'Sleepy Staggers'. It can cause blindness prior to death. Failure of the liver results in ammonia entering the blood and being transferred to the brain, which it destroys.

Another of the questions raised by visitors to the Poison Garden was, 'Isn't it illegal to grow ragwort?' It is a plant that is widely misunderstood. As mentioned in the previous chapter, it was an injurious weed under the Weeds Act of 1959 and was said to cause more economic harm than all other plants put together.

The provisions of the Weeds Act were amended in respect of ragwort by the Ragwort Control Act on 20 November 2003. This Act, which was sponsored by The British Horse Society, originated as a private members' bill, and was presented to Parliament by John Greenway MP. The government gave its backing to the Bill and ensured its successful passage through Parliament. The Act came into force on 20 February 2004.

The Act calls for the creation of a Code of Practice (COP) in respect of ragwort and the code was issued in July 2004. The COP deals with the identification of ragwort, its safe handling, control and disposal, and sets out the duties of livestock owners, forage producers and landowners. One very important point to make is that it is certainly not illegal to have ragwort growing on one's land and it is NOT a requirement to remove ragwort in every case though it is the responsibility of the occupier of the land to take the appropriate action.

The COP (DEFRA 2004) talks about the importance of ragwort to biodiversity and goes on to set out three risk levels:

High Risk:
- Ragwort is present and flowering/seeding within 50m of land used for grazing by horses and other animals or land used for feed/forage production

Medium Risk:
- Ragwort is present within 50m to 100m of land used for grazing by horses and other animals or land used for feed/forage production

Low Risk:
- Ragwort or the land on which it is present is more than 100m from land used for grazing by horses and other animals or land used for feed/forage production

   © Crown copyright 2004 (Crown copyright material is reproduced with the permission of the Controller of HMSO and the Queen's Printer for Scotland)

Immediate action to remove ragwort is only required for high risk areas. For medium and low risk areas, it is a matter of 'wait and see' but with plans being put in place in case the area becomes high risk.

The importance of ragwort to biodiversity can be seen from the number of potential predators the plant has. In normal circumstances these creatures will be able to get what they want from ragwort without causing it significant harm but, as the COP

(DEFRA 2004) points out, some of them in 'plague levels' could completely destroy the plant:

> Biological control is aimed at controlling ragwort by using the plant's natural enemies to lower its density, thereby suppressing ragwort populations and allowing other plants to re-establish. High densities or 'plague levels' of cinnabar moths can destroy complete ragwort populations. Many species feed on ragwort including: cinnabar moth (*Tyria jacobaea*), ragwort flea beetle (*Longitarsus jacobaea*) and ragwort seedfly (*Pegohylemia seneciella*). However their natural spread might not always be as wide-ranging as that of ragwort. Other potential biological control agents include several fungal pathogens (rust diseases). None of these significantly reduces ragwort populations.
> © Crown copyright 2004 (Crown copyright material is reproduced with the permission of the Controller of HMSO and the Queen's Printer for Scotland)

There is no evidence that ragwort causes harm to humans by ingestion, contact or inhalation. Based on known levels required to poison horses, it has been estimated that a human would need to consume 6 kg (14 lb) of ragwort to ingest a lethal dose in one sitting. The only evidence of harm to humans resulting from ragwort, is a couple of cases where very young children were given large amounts of herbal tea made with ragwort which was said to be a cough medicine. Stories of people being killed by eating ragwort or even absorbing it through the skin are all false. Some of those stories provide good examples of how people can misunderstand information or believe erroneous information without questioning it. More than one visitor told of cases they knew when someone had been removing ragwort one day and dead the next day because they had not worn gloves. The skill, which I do not claim to have perfected, was to disabuse them of this false information without causing offence. It does seem that people like to adopt stories and modify them, unconsciously, so that a 'maybe' becomes definite and an unnamed victim becomes 'someone I knew' or even 'a cousin of mine'.

There has been no research done on the effects of handling ragwort, which in itself suggests that this is not a problem, but research on Symphytum spp., comfrey, has shown that absorption of pyrrolizidine alkaloids through the skin of rats resulted in blood levels 20 to 50 times lower than those resulting from ingestion of the same amount. This suggests that a person would need to have, at least, 120 kg of ragwort applied to their skin to achieve the same effect as ingesting the 6 kg referred to above.

Some people cite the Code of Practice as evidence that ragwort can be a poison to humans because, as part of the information it gives on removal, disposal and handling techniques, it calls for the wearing of strong gloves, keeping arms and legs covered. A visitor to the Poison Garden worked as a contractor to local authorities for the removal of ragwort and had never come across anyone experiencing any problems with handling the plant, with or without protective clothing. The COP also calls for the use of face masks to prevent inhalation, which is taken as proof that the poison can be breathed in. It is interesting to note that, in the 2003 draft of the COP, the face mask is 'to reduce the risk of hay fever', but this wording is not present in the final version.

If being poisonous were an evolutionary mechanism to protect a plant what possible merit is there in ragwort continuing to be poisonous when dead? Undoubtedly, the unpleasant taste helps the plant survive and reproduce and is of far more importance from an evolutionary point of view than its being poisonous. Though the mechanisms for causing harm are better understood, ragwort continues to be a nuisance and its presence often enrages people who do not understand how it causes the harm that it does.

Nor is it just a UK problem. Ragwort was introduced in New Zealand in the 1800s. It thrived and became a noxious weed, which farmers sought to eliminate. In the 1930s, sodium chlorate was used as a weed-killer to try and bring ragwort under control. Sodium chlorate is very volatile and fumes penetrated cotton clothing making the clothing flammable. Amongst the incidents reported was the farmer out riding when the friction between his saddle and his clothing caused the sodium chlorate to ignite. In another case, a farmer went straight from work to see his newborn son. When he struck a match to see the child better, his clothes caught fire and he burned to death.

In spite of what has been described as an epidemic of exploding trousers, farmers were so pleased to find something that would control the spread of ragwort that they continued to use it until 1946 when a new type of weed-killer became available.

If evolution has a role in how plants have developed their relationships with other living organisms then taste is, probably, the main result. As we've seen, the horribly sour taste of Senecio jacobaea stops it being eaten before it gets the chance to set and spread its seed.

Knowing yew

Taste is also important with Taxus baccata, the yew. All parts of the yew are poisonous except the flesh of the fruit, which is very sweet but, to my taste, a little slimy. If I were telling this story to a group including children I would, sometimes, look straight at the children and in a stage whisper say, 'If these adults weren't here I'd say it's a bit like eating snot'. The single seed in the fruit is as poisonous as the foliage but, like the apple, the casing of the seed is indigestible. Animals eating the sweet fruit will, therefore, excrete the intact seed, helping the yew to continue its line.

Yew is a plant with a great deal of folklore attached to it and many misconceptions. I always have a chuckle when I hear, or read, that English yew trees were used to make longbows and that's why there are all these old trees in churchyards. Stop to think about it and you'll realise that, if a tree had been used to make longbows, it wouldn't be standing in a churchyard today. Most English longbows were made with imported yew wood because European yew trees tend to grow straighter and with fewer defects leading to a stronger bow.

For me, another key feature of the Poison Garden was to encourage people to question and challenge the things they were inclined to take for granted. I see little difference between someone accepting without question that the yew tree they can see before them was used to make longbows and someone accepting the word of friends that the only way to have a good night out is to drink huge amounts of alcohol and/or snort a few lines of coke.

People would, also, have lots of reasons why yew trees were planted in churchyards without ever realising that yew trees weren't planted in churchyards; churches were built around yew trees. Many of the yew trees you see in today's churchyards are much older than the churches and some are even older than Christianity. The use of yew in graveyards is put down to two reasons. Some say that, because the roots tend to spread out flat rather than going deep, the yew forms a sort of matting layer which the dead, seeking to rise from the grave, are unable to penetrate. The other, more likely, is that the yew has very fine roots which will grow through the eyes of the dead, preventing them seeing their way back into this world.

People often mock Christianity for adopting pagan beliefs but there were often good reasons for so doing. If people want to bury their dead under a yew tree then, by monopolising their availability, you force people to come to church where you can convert them. Also, in the same way that people argue that there was Islam before the Prophet, Christianity didn't begin with the birth of Christ. Associating with older rituals enabled Christianity to say that people had always been Christians but hadn't realised it.

As for the seed, or the stone as some people call it, it is highly poisonous but its outer casing is indigestible. There have been reports of people eating up to forty berries, including the seeds, and coming to no harm because they passed straight through. I've also met someone who said yew berries were his favourite fruit and he ate hundreds every year but, in his case, he spat out the seeds. Equally, there are reports of people eating as few as three berries and being poisoned because they chewed into the seeds.

Though cattle can, and occasionally do, get poisoned by eating yew, this is usually a result of eating dead foliage, yew being another of the few plants that are still toxic when dead. If you have yew in the garden don't ever be tempted to 'recycle' any prunings by dumping them on farmland.

Interestingly, deer are not poisoned by yew. In New England, where many gardens have an unfenced boundary with woodland, gardeners are advised not to plant yew unless they want to be overrun with deer who will, also, eat most of the rest of the garden. I'll be looking at the whole matter of plants being poisonous to specific creatures in Chapter 16. And, like so many poisonous plants, yew may be highly toxic and fatal in fairly small amounts, but there are very few actual cases of poisoning. There are millions of yew bushes around the country, which anyone could munch on if so inclined, but it just doesn't happen.

Paracelsus and the Doctrine of Signatures

In seeking to discover why plants are poisonous the human race is often demonstrating its arrogance. For a long time we've assumed that our race is at the top of the pyramid and everything else in the universe is subservient to us and exists to provide us with some benefit. This led to a belief known as the Doctrine of Signatures. This belief, that the look of a plant told you how to use it, goes back to at least Roman times and is mentioned by Pliny (AD 77–79) but it was not until the 17th century that it was given the name.

WHY ARE PLANTS POISONOUS?

Today, the things people once did based on this belief seem to us to be stupid. Using Lithospermum officinale, gromwell, to treat kidney stones just because the seeds look like little stones is a good example. Pliny's section on this plant indicates just how strongly held this belief was:

> Among all the plants, however, there is none of a more marvellous nature than the lithospermum, sometimes called 'exonychon,' 'diospyron,' or 'heracleos.' ... It bears close to the leaves a sort of fine beard or spike, standing by itself, on the extremity of which there are small white stones, as round as a pearl, about the size of a chick-pea, and as hard as a pebble. ... It is well known that these small stones, taken in doses of one drachma, in white wine, break and expel urinary calculi, and are curative of strangury. Indeed, there is no plant that so instantaneously proclaims, at the mere sight of it, the medicinal purposes for which it was originally intended; the appearance of it, too, is such, that it can be immediately recognized, without the necessity of having recourse to any botanical authority.
> (*Natural History*, Pliny, Book 27. Translated by Bostock J. and Riley H.T., 1856, published Henry G. Bohn)

The Doctrine of Signatures came about because of this idea that everything in the world had been put there by the 'creator' for the use of the human race, and to help us the 'creator' had put these signs on the plants. (Different cultures had different ideas of who the 'creator' was but they shared the arrogant view that everything was for *us*.)

It has to be said that not everyone has abandoned this view. I've even read a supposed authority on cannabis stating that the plant produces substances that are psychoactive to humans entirely because that makes it attractive to humans who spread and protect it. This same arrogance is what leads us to ask 'Why is a plant poisonous?' or, as the question is sometimes framed, 'Why is that plant trying to kill me?' To which the simple and accurate answer is 'It isn't'.

I began this chapter by saying that the answer to why plants are poisonous has been known since the 16th century but, because it doesn't accord with the needs of human arrogance, it was controversial and caused the man who came up with it great trouble. Theophrastus Phillippus Aureolus Bombastus von Hohenheim (1493–1541), who gave himself the single name, Paracelsus (meaning equal to Celsus, a first century Roman revered for the extent of his knowledge), was a Swiss. Depending on the view of his work taken, he is described as a philosopher, aesthete and nutter. It might be argued that the word 'bombast' meaning 'pompous speech or writing' derives from his name rather than the Latin for 'padding', which is the usual explanation.

He is credited with the discovery of hydrogen but, in the plant world, Paracelsus is mostly remembered for two things, one of which was his discussion of the idea of the 'Doctrine of Signatures', though Paracelsus, himself, did not use those words. He was also the first person to say that, in effect, there is no such thing as an absolute poison.

Paracelsus says that every living body contains an alchemist to separate what is useful from what is poison, and states, as fact, that arsenic is removed from a person by passing through the ears. So it would be wrong to claim he had established great truths about the way the human body works but by stating that 'one thing is poison to one and another to another' he does, in effect, lead us to the key that what we call a 'poison' plant means simply that it contains a chemical that is essential to its life cycle, and if that chemical produces a reaction with a chemical that is essential to some creature's life cycle then poisoning occurs. If a creature does not respond to that chemical it will not be poisoned. This explains why birds can eat some things that, to us, are deadly poisons.

So, if you are unfortunate enough to be accidentally poisoned,

WHY ARE PLANTS POISONOUS?

you don't need to worry that there's some evolutionary conspiracy out to get you. You're just one of the, thankfully, few who are unlucky. But it might help your luck to know which plants to be especially careful around.

3

What's the Most Poisonous?

Though usually couched in those words, what many people mean by this question is 'Which plant is the most harmful?' because they associate poisonous with harmful. They are usually surprised to find that some of the most poisonous plants in the world are growing in their own gardens. One couple, on hearing me say that Aconitum napellus, monkshood (or, sometimes, wolfsbane), was one such plant, related how they liked the look of the monkshood so much that they had used it to brighten up their herb garden. Without realising, the wife had included some foliage when she went picking leaves for a salad to go with their meal. The husband went off on a business trip to London so it was only later that they both realised they had suffered severe stomach upsets as a result.

So the most poisonous plants do, sometimes, cause harm but, in general, less poisonous but more attractive plants are the ones that do the harm when it comes to accidental poisoning. And attraction can be visual as with the colourful berries on many poisonous plants or it can be the taste.

Wolfsbane, whaling and warfare

Let's look at Aconitum napellus in more detail. The main alkaloids in the plant are called aconite and aconitine. Of these aconitine is thought to be the key toxin. Ingestion of even a small amount results in severe gastrointestinal upset but it is the cardiac effect, causing a slowing of the heart rate, that is often fatal. The poison may be administered by absorption through the skin or open wounds and there are reports of people being unwell after smelling the flowers. As well as monkshood and wolfsbane it has other common names including mousebane because its smell is said to be poisonous enough to kill mice.

Plants from the Aconitum genus have been used to poison arrow tips, particularly in ancient China. They were also used on arrow shafts in the hope that anyone attempting to remove an arrow from a wounded soldier would get the poison on their hands and absorb it through the skin, especially if their hands had small cuts already incurred in the battle. I haven't found any precise details but it is said that children have died after holding the warm, moist root in their hands.

Its use to poison the tip of a projectile is best seen when applied to harpoons used in whale hunting. The aboriginal peoples of the Kamchatka peninsula, Kurile, Kodiak and Aleutian islands in the far North Pacific Ocean used lances coated with aconite poison to hunt whales. The lance heads were of stone and were intended to break off from the shaft so as to remain in place in the whale's blubber. The whale would be lanced and left to die in the hope that the dead whale would wash up on one of the islands. Each whaler had his own lance head design so that people finding a dead whale would know who had killed it. Cultural rules dictated that the whale belonged to the killer but he would share it with the finders.

The first thing that would be done when a dead whale was found was that the flesh around the wound would be cut out – whether to remove the poisoned area or as a means of retrieving the point of the lance is uncertain. Not all whales would beach in the hunters' islands and some were found by the Nootka people further to the east. It is reported that the Nootka would not eat these whales, though it is not certain whether that was due to the presence of the aconite poison or because, the whale having taken much longer to come to shore, the meat had in any event become rotten.

Because not all whales would end up beached, it is not possible to know how successful the poisoning was, though some whales would have an old set of healed wounds indicating that they had survived a previous poison lance attack.

The exact effect of the poison is, also, unknown. It may be that a big enough dose entered the bloodstream to poison the animal's central nervous system or it may be that it caused only a local irritation, which was enough to cause the animal to thrash around and, finally, die of exhaustion. There are some reports of villagers falling ill after eating whale meat though it seems unlikely that this would be the result of the poison as this is metabolised by the victim, in this case the whale.

One reason for the lack of detailed information is that the use of poison was kept secret in favour of imbuing whaling with magic properties. Thus the fat from the corpse of a dead whaler would be made into grease and applied to the lance head to transfer the skill of the dead whaler to his successor.

In the Kamchatka peninsula and the Kurile Islands, especially amongst the Ainu people, aconite was used to kill other animals as well as whales and also as an arrow poison in warfare. In the Kodiak and Aleutian Islands it was only used for whaling. Anthropologists suggest that this indicates that whaling methods were taught to the people of the Kodiak and Aleutian Islands by the Ainu people of the southern Kamchatka peninsula.

In the 19th century there were attempts to use other plant poisons for whaling. In the 1830s a number of experiments with cyanide were made involving complex harpoons designed to inject prussic acid on penetrating the whale's flesh. The effects were so quick and so deadly that the whalers are reported as being scared to touch the dead flesh. In the 1860s, a Frenchman named Thiercelin experimented with using strychnine and took part in a whaling trip where ten whales were killed with 'bomb' lances containing the poison. All died within 18 minutes.

There is no evidence of any harm coming to anyone who handled these poisoned whales or ate the meat, and adoption of poisoning was only prevented, it is thought, by a realisation amongst whaling crews that such an effective method of killing whales would reduce the number of whaling ships required and damage employment opportunities. This led to them leaking details of what had been secret trials to cause an outcry about the potential of poison whale meat being offered for sale.

So, Aconitum napellus contains highly toxic substances, which, as well as being used when extracted, can cause death in the plant form. That this happens only rarely is due to the taste. It has a characteristic bitter taste that leads to initial burning in the mouth followed by numbness. If swallowed, in even the tiniest amount, several hours of stomach discomfort results. There are a very few cases of fatal poisoning after ingestion of the plant itself, the most recent documented case being a rising young Canadian actor, Andre Noble, who accidentally ate monkshood on 30 July 2004. He was on a camping trip and it is thought that he mistook the root for wild parsnip and included it in his meal. It is, of course, impossible to know if he noticed anything odd about the taste.

William Rhind was a 19th-century Scot who trained and practised as a doctor before turning his attention to studying and writing about natural history in all its forms. In his *History of the Vegetable Kingdom* (1865) he cites a case in Sweden, though without giving a date, where a man exhibited maniacal symptoms after eating fresh leaves of the monkshood. A doctor summoned to assist him expressed the view that the plant was not the cause of his disorder and ate freely of the leaves to prove his point. He died in dreadful agony. We can assume that the haste in which the doctor ate the leaves to demonstrate his superior knowledge prevented him from responding to the taste in time. Rhind, sadly, does not say what became of the patient.

I believe that the taste of the monkshood explains its other well-known use in warfare. It is said to have been used to poison water supplies; sometimes, the water supply to a besieged castle but, more often, the water sources passed by a retreating army. Though this use appears in many books over the centuries there are no reports of huge numbers of casualties as a result. It seems to me that the immediately unpleasant taste of water contaminated with a species of Aconitum would signal that the water had been poisoned and lead the pursuing army to call off their pursuit for fear of being left with no access to safe drinking water.

This recognisable taste has meant that Aconitum napellus and the other species do not make useful murder weapons. The intended victim may notice the taste before a large enough amount has been consumed but, even if a fatal dose has been eaten, comment on the strange taste will rouse suspicion, and the best way to get away with murder is not to have anyone suspect murder has taken place. I only know of one confirmed case where aconitine has been used as a murder weapon and we'll look at that in detail in Chapter 8.

But, just because it isn't a useful murder weapon doesn't mean Aconitum hasn't killed. It was Aconitum lycoctonum, wolfsbane, that was used in some Greek cities to administer the death penalty and, on the island of Ceos, the elderly and infirm were expected to take it to spare their families the burden of looking after them.

That particular story always went down well during tours and I would use one of a number of responses to the reaction of the audience. If one parent/child couple reacted to the story I would ask who had suggested the visit to the Poison Garden. If several couples reacted I would say that, at this point on the tour, it was always possible to pick out the family parties. If there was no obvious reaction, I would look around the group for a few seconds and then say, 'I'm just checking that no one is making notes'. On one occasion, I noticed a little boy, about three or four, drawing on his hand. I couldn't stop myself laughing and had to explain to the whole group what was going on. When we moved on I said to the mother, 'Be very concerned if he asks for a garden voucher rather than a fire engine this Christmas'.

Laburnum, the yellow peril

Sometimes, people think a plant is much higher up the 'most poisonous' chart than it, in fact, is. Laburnum is one such plant. The most usual species grown is the Laburnum anagyroides but this is one plant where the common name and the botanical genus are the same so it's easy to just talk about laburnum.

Laburnum is relevant here for two reasons: first, its actual rather than perceived toxicity; second, because it illustrates the point about the attractiveness of a plant leading to ingestion. All parts of the laburnum are poisonous but most people say it is just the seeds. Though the seeds are the most common cause of accidental poisoning, because they look somewhat like pea pods, poisoning can occur from other parts of the tree. One visitor spoke of losing a kitten after it used a laburnum as a scratching post and then, presumably, ingested plant material from under its claws during grooming. Exposure to the sawdust of laburnum wood has been known to cause 'constitutional symptoms', that is, a general feeling of being unwell but with no specific illness.

In almost every case of laburnum poisoning, reported by visitors, the victim confirmed that they had, previously, been shown how to get raw peas straight from the plant in the vegetable garden and thought the laburnum seed pods were just a different sort of pea. The active ingredient of laburnum is called cytisine, which is similar to, but less strong than, nicotine. Symptoms of severe poisoning are intense sleepiness, vomiting, convulsive, possibly tetanic movements, coma, slight frothing at the mouth and unequally dilated pupils. Children suspected of laburnum poisoning would have their stomachs pumped, and one visitor to the Poison Garden remembered being taken to hospital for that procedure. I asked her if she remembered the effect of the

WHAT'S THE MOST POISONOUS?

poison but all she could recall was her mother slapping her face every couple of minutes on the 45 minute drive to the hospital because, it was believed, it was essential to avoid the onset of coma.

In spite of its reputation, it seems laburnum is not as dangerous as many people believe. A doctor, visiting the Poison Garden, said he had never dealt with a case of laburnum poisoning that resulted in serious illness, and a woman recounted how, as a child, she had eaten laburnum seeds after being told not to and, therefore, did not dare say anything when she experienced a stomach ache. In three days, she fully recovered without her parents realising she was ill and without treatment. Another visitor said they had been researching the subject and had not been able to find a fatal case of child poisoning as a result of eating laburnum.

And, yet, laburnum remains the bogeyman for many people. In recent years, the number of hospital admissions due to suspected laburnum poisoning has dropped to almost none but, in the 1970s, according to R.M. Forrester (1979), there were around 3,000 hospital admissions due to laburnum poisoning each year. Based on the actual toxicity of laburnum, Forrester says: 'It is suggested that laburnum is not as dangerous as has been thought and that many of these admissions are unnecessary.'

It does seem that, for some reason, there was almost public hysteria about laburnum and, I suspect, that some children who did no more than stand too close to a laburnum tree ended up having their stomachs pumped in hospital. I said this to a retired nurse who had worked in a casualty department in the 1970s and her only comment was, 'Do you mean I took all those kicks for nothing?'

Though the belief is less prevalent today people still do hold laburnum to be a serious risk. As recently as 7 June 2007, there was an incident, in Ipswich, involving a laburnum tree and primary school children. The school grounds had been extended by clearing an overgrown area, which put a neighbour's laburnum tree in range. Fifteen children were taken to hospital after being seen playing with the seed pods. There were fears that some may have eaten them but none became ill. The incident occurred during 'Healthy Living Week'.

Many visitors said that, when they moved into a new house, their first action was to remove a splendid laburnum tree 'because of the children (or grandchildren)'. In one case a visitor said her father had decided to cut down a tall laburnum tree when she was three. Having felled the tree, he was called away before he could remove it, meaning the seed pods were now within easy reach for a toddler, and she ate several before being discovered.

Whenever a visitor said they had removed a laburnum tree I would ask them about the other plants in the garden. Some said they had monkshood, many had foxgloves and almost all had daffodils, which, in terms of numbers of incidents, is, in my view, the most harmful plant in the garden.

A host of golden daffodils

The original plant list for the Poison Garden was provided by Caroline Holmes, a writer on herbs and herbalism. When the Poison Garden opened to the public in 2005, quite a few of the plants could hardly be called poisonous and some had been included by Caroline Holmes because of their use as antidotes.

It soon became clear that visitors were more interested in finding out what harm marigolds and fennel could do and that some were going away with the view that this was a herbal garden under a more dramatic name. The Duchess of Northumberland agreed that the garden should be substantially replanted before the 2006 season and the then assistant head gardener, Jane Johnson, came up with a list of plants to add and others to remove, which she and I then edited on the basis of the stories the plants had to offer.

One of the plants that Caroline Holmes hadn't included in her planting was the genus Narcissus, the daffodils. Though I have never come across any record of a fatality, all the plants in the Narcissus genus are toxic if ingested and will cause a stomach upset of some form. Many gardeners lift their daffodils each year, storing them in the garden shed or garage. When I first started researching the stories of the daffodil, to see if it should be included in the replanting, I read many references to the bulbs being mistaken for onions. This seemed rather improbable and, when the daffodils first started showing in spring 2006, I would state this and try and move on quickly. But visitors kept saying, 'Yes, I've done that'.

At first, I tried to keep track of every story told by visitors about their experience with daffodils, and other spring bulbs, but over the three springs I was in the garden with the daffodils there were so many that I lost count. What follows is just a selection of some of them.

One of the earliest stories was from a visitor who poisoned herself and her dinner party guests by grabbing daffodil bulbs rather than onions when she didn't bother to switch the lights on in the garage. In another case, an *au pair* was left to prepare the family meal and used daffodil bulbs instead of onions.

A visitor, certainly aged over 60, related how at 12 years old when her mother was 'away' she cooked a stew for the family. She went to get some onions from the garden shed where they were kept but selected daffodil bulbs instead. A man, probably in his seventies, related that his mother, when a child, had been poisoned when the family's maid put daffodil bulbs in the meal.

An example of the effect of a specific amount of toxin came when a couple related their experience of daffodil poisoning after an ageing aunt included bulbs instead of onions in a meal. The husband ate a whole bulb whereas his wife ate only half. He was violently sick after ten minutes and then recovered fully. She was not taken ill for an hour but then suffered two days of nausea, vomiting and diarrhoea.

Another visitor said, 'I know about them, I fed them to my parents.' When he was quite a young child, both his parents were ill in bed so told him how to make a stew for them but not how to distinguish between onions and daffodil bulbs.

Incidentally, though the bulbs are the most frequent cause of poisoning, the rest of the plant contains toxins. In 2003, a case was reported by the London Centre of the National Poisons Information Service in which foliage from daffodils had been taken for leeks and included in a meal, causing stomach upsets for the three people who consumed them.

It's not just daffodils which are toxic. Galanthus nivalis, snowdrop, has similar effects. A man recalled being told to prepare the evening meal while his wife was out. When she returned he happened to remark that he'd 'used those small onions'. Fortunately, his wife knew she had no onions so he must have used the snowdrop bulbs, which were awaiting planting. The meal was consigned to the bin and the couple had bread and jam for tea. What makes that story of particular interest is that the man was a farmer so, if a farmer can't distinguish between a snowdrop bulb and a shallot, there is little hope for the rest of us.

Even tulip bulbs are poisonous, though many people dispute this because they know that, during the 1944/45 'Hunger Winter' in the Netherlands starving people made bread using a flour from crushed tulip bulbs. It was found that removing the centre of the bulb, as well as the skin, removed enough of the toxin to enable the rest to be dried and ground to make a sort of bread flour.

What happened during that terrible winter over sixty years ago is still being studied and seems to have lessons for us even today. Researchers in the Netherlands have now studied people whose mothers were pregnant during the period mid-October 1944 to mid-May 1945, the Dutch Hunger Winter. They found a statistically significant link between people whose mothers were in the first three months of pregnancy in this period and later mental illness. There was no significant link later in pregnancy. The study confirms earlier suggestions that malnutrition in early pregnancy inhibits proper brain development. The researchers raised concerns about future mental health problems in parts of the world affected by serious famine.

But my favourite story about ingesting bulbs was not a matter of accidental poisoning. A recently qualified dentist said that one of her fellow students was convinced he was going to fail an examination, which would have led to him having to leave the university. Before the exam started he ate a daffodil bulb and began vomiting after half an hour, forcing him to leave the exam room. Because of this he was allowed to resit the exam at a later date when he was better prepared.

So the answer to 'What is the most poisonous plant?' depends on whether you mean the one that is theoretically most poisonous, or the one that causes the worst effects for the few people who accidentally ingest it, or the one that harms the most people, albeit only causing an unpleasant stomach upset. But it's quite probable that the expected answer can only come from asking the question a different way. In the next two chapters we'll look at which plants actually are the biggest killers even if they are not the most poisonous.

4

What is the Biggest Killer? Part 1

On the gates of the Poison Garden is the message 'These Plants Can Kill'. Note, 'Can Kill' rather than 'Do Kill' because, without the intervention of a living organism, the plants are incapable of killing. The greatest harm, of course, comes when that living organism is a human being.

So, you can't blame the plants that kill the most for achieving this dubious status. They only become killers when the human race decides to make them so. If we could change our ways they would no longer be killers. Take the genus Helleborus, the hellebores. The roots of the hellebores are powerfully emetic and up until the 17th century, children were given hellebore root to attempt to clear them of the internal worms that were the frequent result of eating rotten meat. The trouble is our ancestors didn't understand that there are different types of worm and they don't all live in the stomach so they would often give the child a fatal overdose while trying to remove a worm that was firmly ensconced in the intestines. An acceptance of high infant mortality rates and the lack of any post-mortem investigations and records mean we have no way of knowing how many children died in this way. Today, the incidence of worms is greatly reduced and we know better than to use hellebores medicinally.

There are three ways of being killed by a poison plant. You can eat it, accidentally, thinking it is something else but this leads to very few deaths each year. In fact, so few that the EU doesn't even bother collecting the numbers of fatal plant poisonings separately. The USA does and figures from the USA suggest that the average is around five deaths a year.

The second way a plant substance can kill is if someone gives it to you in order to kill you. Statistics for murder using plant-based poisons are not susceptible to averaging because, of course, the discovery of a serial killer, like Charles Cullen or Harold Shipman, causes a spike in one year's figures and the statistics only include known incidents. But, again, the overall numbers are in the tens per year because, these days, there are plenty of other ways to kill someone rather than relying on a plant poison.

And then there is the third way, where a person self-administers a plant poison because they want the psychoactive effect it provides, and either ignores or misjudges the dangers associated with it or becomes addicted to it and is unable to prevent its fatal consequences; in other words, the 'drugs'.

A not so pleasant comedy

Some deaths, in this category, do occur from direct use of plant material often by boiling leaves in water to make a 'tea'. These days the genus Datura has been re-evaluated and divided into Datura and Brugmansia though most of the stories about the different species are from the days when Datura was used for them all. These two genera seem to be the biggest killers as far as direct use of a plant is concerned.

Datura stramonium and Brugmansia sauveolens are two of the most commonly mentioned species when poisoning following use as a drug has occurred. Datura stramonium is the 'Jimsonweed' (Jamestown weed), which acquired this name in Jamestown, in 1679, when soldiers ate leaves in a salad and experienced 'a very pleasant comedy'.

> Some of them eat plentifully of it, the Effect of which was a very pleasant Comedy; for they turn'd natural Fools upon

WHAT IS THE BIGGEST KILLER? PART 1

it for several Days: One would blow up a Feather in the Air; another would dart Straws at it with much Fury; another, stark naked, was sitting in a Corner, like a Monkey, grinning and making mows at them; a Fourth would fondly kiss and paw his Companions and snear in their Faces with a Countenance more antick than any Dutch Droll ... A thousand such simple Tricks they play'd, and after Eleven Days, return'd to themselves again, not remembering anything that had pass'd. (Beverly, 1705)

Brugmansia sauveolens is 'angel's trumpet', so called because of the shape of the flowers though I would tell visitors it was because if they had enough of it the angel's trumpet would be the next sound they heard.
Far from the 'very pleasant comedy' described above, the use of these plants, in overdose, is reported in case studies to produce confusion, delirium and hallucinations with drowsiness, sleep or coma generally following. Dilation of the pupils is such a common effect it gets mentioned only in passing in some reports. Agitation and convulsions requiring the use of restraints or sedatives are reported in around a third of the cases. It is also around a third of the cases where death is given as the outcome of Datura poisoning.
That said, I have only been able to find information on a total of 267 cases of poisoning including 188 in a six-year period in Texas, so, clearly, the total number of fatalities each year due to the Datura/Brugmansia genera are few.

Drugs kill!

To start ramping up the body count we need to start looking at substances that have been highly processed so that their plant origins are no longer obvious. This usually leads people to think of substances like cannabis, cocaine and heroin because as we all know 'Drugs kill!'

IS THAT CAT DEAD?

Actually, cannabis does not kill. There is only one case where an overdose of cannabis has been cited by a coroner as a cause of death and, even there, it was assumed that the deceased must have ingested a large amount of cannabis because smoking could not result in a fatal overdose. As we'll find in Chapter 13, the deaths attributed to being under the influence of cannabis are often due to other factors.

Cocaine certainly does cause deaths though many of these are not the result of chemical effects due to its use. Cocaine is the main alkaloid found in the Erythroxylum coca plant and is one of the most powerful stimulants known to the human race. Cocaine's effect is almost immediate and, depending on the dose and the method of administration, may last from 20 minutes to a few hours. It produces euphoria with restlessness and hyperactivity but can also increase blood pressure and accelerate the heart rate, potentially to dangerously high levels. Regular use over a long period substantially increases the risk of cardiac problems as a result of the elevated heart rate. High blood pressure can also be a longer term problem. Collapse of the septum, the tissue between the nostrils, is a well-known effect of regular use.

Short-term, most of the harm done by cocaine is associated with hyperthermia, the elevation of body temperature, which can produce kidney failure. There can be added risk of death if cocaine is consumed with alcohol since the metabolite formed, cocaethylene, has been shown to be the most active of the metabolites of cocaine. One visitor to the Poison Garden spoke of losing her son to a night of binge drinking and snorting coke.

In 2004, 149 death certificates issued in England and Wales mentioned cocaine overdose as at least a contributory factor in the cause of death but it is hard to know how many of those deaths were entirely due to cocaine.

To a large extent the biggest cause of death associated with cocaine is the number of murders occurring in producing areas, especially Columbia. Recreational cocaine use is often seen, by users, as a victimless crime but the deputy president of Columbia

WHAT IS THE BIGGEST KILLER? PART 1

said, in 2007, that every dose of cocaine consumed in a nightclub in Europe or the USA provided fuel for the murder of Columbian citizens.

It's not surprising that, when planning the Poison Garden, the Duchess of Northumberland was anxious to have the Erythroxylum coca plant and a great deal of effort went into getting a licence from the UK Home Office to possess the plant for educational purposes. Unfortunately, very little effort went into obtaining a plant or considering whether it would survive outdoors in the north-east of England.

When the Poison Garden first opened there was no coca plant, and no one seemed to be doing anything to try and obtain one. The problem is that there is a UN convention forbidding the transfer of the plant or its components across international boundaries except in two particular circumstances. Cocaine is still used as a pharmaceutical, though to a lesser extent than when it was used as a dental anaesthetic, and trade for that use is permitted. The other exception permits the shipment of leaves from which the alkaloids have been removed for use as food flavouring. Though the Coca-Cola company never makes public the contents of its drink, many people believe that this exception is because, even though cocaine itself was removed from the recipe in 1901, Coca-Cola is still flavoured with coca leaves. Certainly, export statistics show that almost all denatured coca leaves are supplied to the USA. According to the Internet news service, Ananova, two German scientists, Udo Pollmer and Susanne Warmuth, in a book whose title translates as *Encyclopaedia of Popular Diet Myths*, say they have completely broken down the content of Coca-Cola and found that the aroma is the result of coca leaves, lime, coffee and cocoa distillations, mandarin leaf tincture and a little ginger.

Anxious to be able to show visitors what the Erythroxylum coca plant looks like, I began to search for a way to obtain one. Eventually, after much digging I found someone who would supply seeds of the coca plant. Though the supplier was in Europe, the seeds were sent direct from South America and were in a poor condition on arrival and, therefore, failed to germinate. The supplier counselled a 'wait and see' approach but, eventually, accepted that the batch of seeds were duds and supplied a further batch. Of the 15 seeds supplied in this second batch, 2 germinated and began to grow in the Alnwick Garden greenhouse.

By now, this was the autumn of 2007 and we hoped that we might have a plant capable of being displayed during the summer of 2008. Unfortunately, something happened during the early months of 2008 and both plants died. By this time, I had built up a rapport with Dr Liz Dauncey, the poison plant expert at Kew Gardens, and felt bold enough to enquire if we might be able to get cuttings from their plant. Liz replied that it would not be possible to provide cuttings but they had a mature plant, which they were intending to dispose of, and we were welcome to have that and take our own cuttings.

It now just became a matter of how to transfer the plant from one to the other. Kew, naturally, insisted that I go and collect the plant in person and contacted the Home Office to see what paperwork would be required. After several weeks of dancing around the issue, the official at the Home Office acknowledged that there was no procedure for transferring such a plant from one licensed holder to another and suggested that we just get on and make the transfer.

So it was that early one Monday morning in May, having driven down the previous afternoon, I presented myself at Kew Gardens and just about managed to fit a very large Erythroxylum coca plant in the back of my Freelander. Nigel Rothwell, who Liz had directed me to for the transfer, had the Kew paperwork ready but said that it needed to be signed by his boss who was not in that morning so he would send our copy in the post. Thus, I set off with no paperwork of any sort explaining why I had what, in effect, was a Class A substance in my car. I had deliberately arranged it that even the very few people at the Alnwick Garden who knew the plant was coming did not know when, so I would have been pretty much on my own if for any reason a policeman had asked me what I had in the car.

WHAT IS THE BIGGEST KILLER? PART 1

I like to think I'm a pretty safe driver but I certainly paid extra attention that day and only let the vehicle out of my sight for the shortest of 'comfort stops' halfway up the M1.

That was my last act as Poison Garden Warden. I understand that cuttings were taken from the plant before it was put on display during the summer of 2008. As expected, it did not fare well and when I visited the Poison Garden, in September, it was looking quite poorly. In 2009, no coca plant was on display but I don't know if the new plants were not substantial enough to bring out or if the cuttings failed.

The extent to which cocaine causes harm is still the subject of debate. In 1995, the World Health Organisation, WHO, was prevented from publishing details of the largest survey ever undertaken of cocaine use. This survey found that for the overwhelming majority of users of cocaine in all its forms there was no evidence of any harm. Indeed, many people argue that chewing the leaves in South America is akin to the drinking of tea and coffee in the rest of the world. In a small minority of users, however, harm resulting from dependence does arise but it is unclear whether this is true addiction.

Stories from the past of the Erythroxylum coca rely on the exploitation of that dependence/addiction which, for ease, I will refer to as addiction for the rest of this book. And those stories help to make a central point about the trade in illicit substances.

The word 'coca' comes from the Aymara word q'oka, which means 'food for travellers and workers'. The addiction potential of cocaine was well known by Peruvian Indian kings. Their kingdoms were widely spread across the mountains and their only hope of keeping control was by being able to quickly send and receive messages from the outlying settlements. To do this they used messengers who were given coca leaves to chew. Cocaine is a powerful appetite suppressant as well as a stimulant so the messengers could cross the barren high mountain passes without feeling hunger. For the kings, the addictive properties of the coca were a positive advantage as they could be sure that the messengers would travel quickly and would deliver their messages as this was the only way to obtain further supplies of coca leaves. Thus, the kings knew they were controlling their kingdoms and the long-term health of the messengers was of no concern because they could always recruit more.

In the 16th century, the Spanish set up silver mines with slave labour given coca leaves to chew to keep them working in the most appalling and unsafe conditions. It was useful that it was addictive because they kept coming to the mines to get coca leaves to chew. So both the kings and the Spanish got what they wanted and didn't care about the harm it caused to the people they gave it to.

Though I have only seen the story in one publication and have not been able to confirm it, it has been said that, at the start of World War I, the German army command had a plan to give their soldiers a daily ration of cocaine to both stimulate them to greater efforts and reduce the logistical problem of feeding an army spread across Europe, cocaine being an efficient appetite suppressant. It was only the realisation that they could not source enough cocaine and would have an entire army suffering withdrawal effects that, so the story goes, changed their minds.

So, throughout its history, coca leaves and cocaine have been supplied for the benefit of the supplier with little concern for the effect on the user. And that remains the case today with all illicit substances, not just cocaine.

Drugs kill

So if cannabis and cocaine are not big killers, what about heroin? What we call heroin is actually diamorphine, a chemical derived from processing morphine, the principal alkaloid found in Papaver somniferum, the opium poppy. The earliest reference to opium comes from ancient Egypt. A papyrus, dated to about 1500 BC but believed to be recording long-known remedies, suggests opium as

WHAT IS THE BIGGEST KILLER? PART 1

a means of getting small children to sleep. For this purpose it was mixed with fly excrement, which suggests there must have been a lot of flies in Egypt if it was practicable to collect their dung! It also makes me wonder if the soporific effects of the mixture were more psychosomatic than physical. Certainly, I think the second time my mother said to me 'If you don't go to sleep I'll give you some more of that opium and fly crap' I'd be asleep pretty quickly.

Addiction to opium seems to have come about more from its use as a painkiller rather than desire for recreational effects. The popular image of the Chinese opium 'fiend' smoking opium in a sordid den ignores the fact that, in 18th- and 19th-century China, there might be 100,000 people for each medical practitioner. In those circumstances, people had to rely on their own remedies and opium proved to be a most useful one with just one disadvantage; it does produce addiction.

Whilst China's preferred method of using opium was to smoke it, in Europe opium was 'eaten' in a preparation known as laudanum, which was a tincture of opium in alcohol. Perhaps the most famous opium addict of the 19th century was Thomas De Quincey, whose *Confessions of an English Opium-Eater* is regarded as a literary classic as well as a detailed account of the effects of opium. But, the most telling effect of opium is seen not in the actual text but by a simple comparison of the first edition with the second edition which was published almost 30 years later.

In the first edition, De Quincey writes about giving opium up and the hope he has for the future. By the second edition he has been an opium addict for many years and has long abandoned any hope of freeing himself from the addiction. The spare, considered language of the first edition has gone and the second edition, which is almost three times as long, has the rambling quality of the hopeless addict seeking to justify their actions but unable to discipline their thoughts to make a cogent argument.

By the early 19th century, there was growing concern at the effects of opium addiction and scientists began to look at the substance in some detail. In about 1817 (sources vary as to the exact date), morphine was isolated from opium. Initially, it was thought that morphine could be used to cure opium addiction; the knowledge that morphine was the cause of addiction would not come until later.

Morphine addiction replaced opium addiction in the developed world and, in 1874, diamorphine was first synthesised by C.R. Alder Wright and it is this form, known as heroin, which gives us the most harmful of the illicit drugs.

It is worth pointing out that Thomas De Quincey was an opium addict for the vast majority of his life and was 74 years old when he died so early death as a result of heroin use is by no means certain. It does, however, cause a significant number of deaths, directly and indirectly.

Heroin is addictive in the true sense of the word. That is to say users develop a tolerance for it and need ever larger doses to achieve the same effect. And they suffer physical withdrawal symptoms if they are not able to obtain sufficient supplies.

The development of tolerance means that it is impossible to specify what is a fatal dose. What would kill someone who had never used the drug before would be a fraction of the daily dose for an addict. An interesting example of this comes from an unexpected place. In *Edison, His Life and Inventions* (Dyer and Commerford Martin, 1910), the authors provide a substantial extract from Edison's own notes about a strange visitor to the Menlo Park (New Jersey) workshops where Edison conducted most of his work.

> At Menlo Park one cold winter night there came into the laboratory a strange man in a most pitiful condition... He said he was suffering very much, and asked if I had any morphine ... so I got the morphine sulphate. He poured out enough to kill two men... He said he had taken it for years, and it required a big dose to have any effect. I let him go ahead. In a short while he seemed like another man and began to tell stories... [H]e finished every combination

of morphine with an acid that I had... Then he asked if he could have strychnine. I had an ounce of the sulphate. He took enough to kill a horse, and asserted it had as good an effect as morphine. When this was gone, the only thing I had left was a chunk of crude opium... He chewed this up and disappeared.

It seems to be that misjudging the state of this tolerance is one of the main causes of death from heroin, which, typically, comes from respiratory failure as a result of the excessive depression of the central nervous system. Death from heroin, it appears, may most often occur when an addict has been trying to kick the habit and relapses without realising that his or her tolerance has decreased or may result from use of unexpectedly strong heroin. Most heroin on the street is 'cut' with other materials to make it go further. Buying from a different source can lead to an addict obtaining material of a higher strength.

Perhaps the most telling demonstration of the effects of developing tolerance is seen in the number of deaths of released prisoners. Though drugs are undoubtedly available in British prisons, the amounts available are restricted. Thus imprisoned addicts simply get by while incarcerated and, often, do not realise that their decreased use results in a decrease in tolerance. Upon release, they go straight back to their pre-imprisonment levels of consumption and die from overdose. A recent study found that men are 29 times, and women 69 times, more likely to die in the first two weeks after release from prison than the general population, and many of these deaths are attributable to addicts thinking they can go back to their pre-incarceration levels of consumption.

So, heroin certainly does kill but determining the number of deaths is more complex. In addition to deaths resulting directly from the effects of the substance there are deaths attributable to the way of life of many drug addicts. Becoming HIV positive as a result of shared needle use may result in death from AIDS, which could be attributed to heroin, as could death from malnutrition or many other causes of death for those whose lives are spent, mostly, on the street. Quite a large number of heroin addicts will also abuse other substances so the decision to call a death heroin related can require a fine judgement of the relative effects of different substances.

Overall, the United Nations Office on Drugs and Crime (UNODC) believes around 200,000 deaths a year are attributable to what it calls 'problem' drug use and the overwhelming majority of these will be due to heroin. In Europe, statistics show that somewhere between 7,000 and 8,000 deaths a year are directly the result of a heroin overdose. Whatever the 'true' figure for deaths from heroin, if there even is such a figure, it is clear that we have taken a substantial step up from the number of deaths attributable to cocaine.

So, as already mentioned, heroin is 'the most harmful of the illicit drugs'. But where does that place it in terms of all harmful substances? In the next chapter, we'll look at two substances that make the harm caused by heroin appear very minor.

5

What is the Biggest Killer? Part 2

Whenever I took a tour around the Alnwick Garden Poison Garden, I would try and make each story a little drama building up to a climax. The training sessions I ran for other staff and volunteers all stressed that the 'This plant is X. It does Y' type of guided tour was not what we were aiming for. Of course, my approach did occasionally lead to misunderstandings.

'Now, we come to the biggest killer of them all. The plant that, every year, kills more people than all the other plants in this garden have ever killed in their entire history. Nicotiana, the tobacco plant.' And, sometimes a visitor would say 'How?'

Even after the explanation there would be some who didn't think smoking counted as poisoning. I'll return to tobacco later in this chapter but, first, I want to look at the second position holder in the 'biggest killer' stakes.

So, tobacco is the biggest killer of them all, but between heroin and tobacco is another substance causing a great many deaths as well as many non-fatal conditions and psychological problems for its users: alcohol.

When the replanting of the Poison Garden was being discussed, the question of whether we should include a plant such as Humulus lupulus, the hop plant, so as to bring in the subject of alcohol arose. In the end it was decided that the presence of Artemisia absinthium, the wormwood that was used to flavour absinthe, would provide enough opportunity to discuss alcohol with groups, especially school groups, where it was appropriate.

The Green Fairy flies again

That was not such a bad decision because the balance of opinion seems to be that the harm done by absinthe was more a result of its being a strong alcohol than anything specifically related to the Artemisia absinthium.

Trying to get at what the true effects of absinthe drinking in the first three quarters of the 19th century were is typical of trying to understand the effects of many psychoactive substances. So, for that reason, it is worth looking at the topic in some detail.

The 'evidence' presented for a particular effect is likely to be coloured by the point of view of the presenter. Thus, it may well be true that the campaign against absinthe in the late 19th century was biased by an unholy alliance of prohibitionists with the French wine industry, which was anxious to recover the market share lost during the outbreak of grape phylloxera, which began in the middle of the century. For the wine growers, it was important to find grounds for attacking absinthe, which did not rely on the inherent harm potential in all alcohol consumption.

It seems to be the case that the majority of those saying that absinthe is not the demon it was made out to be either have a financial interest in absinthe sales or are avid absinthe drinkers. Then there are those who want absinthe to appear to be capable of delivering a 'legal' high so as to sell it to people who would not consider using cannabis because of its status.

Both these groups want to demonstrate that absinthe today is no different from what the great artists are alleged to have drunk in France.

But, absinthe's bad reputation dates from long before the end of the 19th century so is it possible, from this distance in time, to say if there is any truth in its alleged ability to cause hallucinations and convulsions and the condition described in the 1860s as 'absinthe epilepsy'?

The suggestion that the effect of thujone, the active component of Artemisia absinthium and, thus, absinthe, results from its similarity to delta–9-tetrahydrocannabinol (THC), the principal active ingredient of cannabis, comes from an article, published in *Nature* (Anderson, Del Castillo and Rubottom, 1975). The authors started from the point of view that absinthe and cannabis produced similar psychological effects and hypothesised that this resulted from similarities in the molecular structure of thujone and THC, which might result in the substances binding to the same receptors in the central nervous system.

It should be stressed that the authors '*propose* [my italics] therefore that both thujone and THC exert their psychotomimetic effects by interacting with a common receptor in the central nervous system' and conclude that 'This hypothesis suggests new experimental approaches'. It must also be remembered that this was a short article and not a full scientific paper.

Nonetheless, this association between thujone and THC was seized on and, quickly, became the accepted wisdom. Note that in an article in *Trends In Pharmacological Sciences*, B. Max (1992) states 'The literature on the pharmacology of thujone is, to put it bluntly, second rate', so there was little work going on to challenge this position.

The rapid uptake of this erroneous view may say something about the debate on cannabis. 'Everybody knows' that 19th-century absinthe was a very dangerous substance so demonstrating that this was because it behaved like cannabis might help to reinforce the message on the alleged harm caused by the latter.

It may also have to do with companies projecting the image of cannabis onto their versions of absinthe. Logan Distribution, Inc. on its website repeats the claimed similarity between thujone and THC and goes on to state that the absinthe it offers, in the USA, is 'as strong as the legitimate Absinthes of the 19th century'. That claim is worthy of further examination.

The true nature of the action of thujone was described in a paper by Casida, Höld, Ikeda, Narahashi and Sirisoma (2000). This says that thujone can greatly disrupt the nervous system and damage the brain's self-control mechanism producing epilectiform convulsions, hallucinations and delirium. But, did mid-19th-century absinthe contain enough thujone to produce these effects?

As with any substance, the quantity consumed must be expected to have a bearing on the effect produced. The claimed effects of 19th-century absinthe depend on the perception that it had much higher thujone levels than modern versions.

Professor Wilfred N. Arnold, in his book *Vincent Van Gogh: Chemicals, Crises and Creativity*, quotes a typical thujone level in 19th-century absinthe of 260 mg/l. It has not been possible to find the work that gives this level but it is believed to come from replicating old recipes for absinthe and measuring, or calculating, the thujone level. These postulated levels have been challenged, most recently in the *Journal of Agricultural and Food Chemistry* (Breaux, Kuballa, Nathan-Maister, Lachenmeier, Schoeberl, and Sohnius, 2008).

It should be pointed out that two of the authors have a financial interest in sales of absinthe, and all the 'pre-ban', that is pre–1915, samples tested dated from 1895 at the earliest. Vincent Van Gogh, said to be one of the most prominent absinthe drinkers, died in 1890. By the last decade of the century, grape phylloxera, first seen in France in 1863, had caused considerable damage to the French wine industry and it is possible that absinthe makers revised their formulations to encourage more wine drinkers to sample the spirit.

It is believed that Professor Arnold based his work on recipes found in Pierre Duplais's 1855 book whose title translates as *A Treatise On The Manufacture and Distillation of Alcoholic Liquors*. This might explain the difference between Professor Arnold's thujone levels and those found in turn of the century products.

Reports, talking about these results, speak of the European Union lifting its ban on absinthe without stating that the thujone levels are required to be below 10 mg/l. Again, it seems, modern day suppliers of absinthe are seeking to demonstrate that their wares are 'the real thing' by claiming similar levels of thujone to 19th-century versions.

WHAT IS THE BIGGEST KILLER? PART 2

As recently as December 2007, writing in *Scientific American*, Professor Arnold said that thujone 'can cause hallucinations, convulsions and permanent damage to the nervous system'.

Lachenmeier and his colleagues (2008), however, try and suggest that absinthe is no more harmful than any other strong alcohol and that it is the alcohol that is capable of causing harm and not the thujone or other herb extracts used in the drink.

There is some suggestion that deliberate adulteration or accidental contamination might have resulted in the production of absinthes that did cause the effects ascribed to thujone but it is also said that many of the effects seen were simply those of acute alcoholism.

The term 'absinthe epilepsy' appears to have been coined by Dr Magnan who, in the 1860s, had some 250 absinthe abusers in his care and based his new term both on his observations of these patients and by experimentation on animals. In 1895, the Royal Society published a paper detailing experiments on cats to determine the progress of absinthe epilepsy. In a paper delivered to a conference on eugenics in 1912, Dr Magnan describes in detail the attacks suffered by absinthe abusers and says 'it is exactly like an attack of epilepsy' ('Alcholism and Degeneracy' Problems in Eugenics. Papers communicated to the First International Eugenics Congress held at The University of London, 24–30 July 1912. Alcoholism and Degeneracy by M.M. Magnan and A. Fillassier.)

There is no dispute about the harm that can be caused by the oil of wormwood extracted from Artemisia absinthium which contains high concentrations of thujone, but it seems unlikely that there will ever be a consensus on whether there was sufficient thujone, even in crudely produced absinthe, to carry those effects into absinthe drunk in 'normal' quantities.

The debate over the harm or otherwise of absinthe in many ways typifies the debate over the harm caused by alcohol in general. Alcohol is, probably, the world's most widely used psychoactive substance. I say 'probably' because it is hard to know if more of the world's population drinks alcohol than smokes. In non-Muslim communities, especially in the 'West', the rate of alcohol use certainly exceeds the rate of smoking but whether that remains the case once the teetotal Muslim population is included is hard to say.

One too many

The alcohol-using people of the world can be divided into four groups. There are those for whom their consumption causes no short- or long-term problems because the amount they consume both on a day-to-day basis and throughout their lives is too small to produce any noticeable effect. There are those who suffer the occasional short-term problem because their consumption at a single event exceeds their body's ability to process the poison. These are the people who have 'one too many' at the company Christmas party or a relative's wedding. The third group are those who have a high average level of consumption with regular spikes of 'binge drinking'. This group is liable to suffer longer-term problems with the liver, pancreas and other organs. Then there is the fourth group whose drinking leads to addiction and, if the addiction remains untreated, often leads to premature death.

Estimates of worldwide alcohol related premature deaths vary, with The Institute of Alcohol Studies estimating that alcohol is responsible for five times as many deaths as illicit drugs (a figure of one million), and the World Health Organisation (WHO) saying just over two millions deaths have alcohol use as a factor.

But, whether it is five times the rate of illicit drugs or ten, it is clear that alcohol is a much greater cause of harm and death than the illegal substances that are grouped together in the mantra 'drugs kill'.

But, even if the WHO estimate is closest to the true figure for premature deaths due to alcohol, that number pales when seen against the estimated five million people a year who have their lives shortened by their addiction to tobacco.

WHAT IS THE BIGGEST KILLER? PART 2

Up in smoke

Nicotiana tabacum is the species primarily used to produce tobacco commercially though there are parts of South America where Nicotiana sylvestris, the species which will grow as a summer annual in the UK, is used.

It was Leonhart Fuchs (1501–1566) who coined the name 'Nicotiana' after Jean Nicot who sent seeds of the plant to Francois II and the French court c.1559. Nicot's credit as the first to bring the plant to Europe is wrong as it was known in the Low Countries after being brought there by Spanish merchants in the 1540s. Knowledge of the plant by Europeans dates from 1492 when Columbus's sailors saw it being smoked in Cuba and Haiti.

Fuchs is scathing of the many who believed the tobacco plant to be a variety of henbane. Had his encyclopaedia of plants been published it would have contained very detailed illustrations of all parts of the plant to demonstrate that it was not related. Even so, John Gerard, in 1598, calls it 'Henbane of Peru' and begins by comparing its effects to henbane. His descriptions of its virtues cover two full pages. The following is a selection of what Gerard has to say:

> A remedy for the paine in the head called Migram or Migraime that hath been of long continuance...
>
> It is good against poison, and taketh away the malignity thereof...
>
> The oile or juice dropped into the eares is good against deafnesse...
>
> Many noticeable medicines are made hereof ... which if I were to set down at large, would require a peculiar [that is separate] volume...
>
> It is also given to such as are accustomed to swoune, and are troubled with the colicke and windinesse, against the dropsie, the wormes in children, the piles and the sciatica...

IS THAT CAT DEAD?

It is, of course, the tars in tobacco that have been found to be the main harmful ingredients, but nicotine, the principal alkaloid, is a dangerous poison in its own right. Symptoms of nicotine poisoning are said to be loss of motor control, involuntary evacuations and convulsions prior to death. It can be absorbed through the skin and it is often said that when tobacco was first brought from America sailors would tie leaves to their bodies to smuggle them but would die from absorbing the nicotine when sweating on the voyage home. Peter Macinnis (2004) in *Poisons – from Hemlock to Botox and the Killer Bean of Calabar* says there was just one case, in the 19th century, of such poisoning but says that in 1970 problems caused by tobacco pickers in North Carolina putting picked leaves under their arms were first identified as an industrial disease. As we shall see, in Chapter 9, the armpit has been used before as a means of rapidly absorbing poison.

The spring 2004 edition of *Poisons Quarterly*, the regional newsletter from the London Centre of the National Poisons Information Service, cites the case of an eleven-month-old boy who was noted to be restless following his morning bath. He vomited three times after approximately 90 minutes. Shortly after, he experienced a seizure lasting one minute. In hospital, he was observed for an hour with little change and then his nappy was removed. A nicotine patch was discovered stuck to his bottom. Within fifteen minutes of its removal his condition improved and he was fully recovered after four hours. The discarded nicotine patch had missed the bin and landed on the bathroom floor where the naked child sat on it after his bath.

A researcher for Philip Morris, the tobacco company, committed suicide in 1982 by drinking liquid nicotine. Liquid nicotine is used in the manufacture of products like patches and chewing gum intended to reduce smoking. There has recently been speculation about her choice of suicide method with claims that her work wouldn't have brought her into contact with liquid nicotine so she must have gone to great lengths to obtain it from another part of the building.

WHAT IS THE BIGGEST KILLER? PART 2

The speed with which nicotine both kills and metabolises so that it cannot be found in the corpse has led to it being a favourite poison in works of fiction. In an episode of the American crime series *CSI*, the victim was murdered with liquid nicotine, which was added to her cherry brandy. Liquid nicotine metabolises quickly and is supposed to be almost undetectable if the victim is a smoker. In this instance, however, the convulsions produced resulted in the victim throwing herself through a window and bleeding to death so that a measurable amount of nicotine was still present in her body. It would seem that the writers took their inspiration from Truman Capote's novella, *Hand Carved Coffins*, which also features a murder with liquid nicotine. This may also have inspired the writers of *Midsomer Murders* where, in episode seven, an *au pair* was run over by a car and then injected with liquid nicotine.

But, though nicotine is a deadly poison with records of causing death both in fact and fiction, it is smoking that elevates its position in the league table of harm. From its earliest introduction, smoking has been a controversial practice with those in authority seeking to control or outlaw it.

James I's 1604 'A Counterblaste to Tobacco' is often cited as an example of efforts to control its spread but there are those who suggest that what worried James more was that smoking as a social activity was being used as a cover for groups of opponents to his rule to gather and discuss subversion. It is also possible that he was responding to pressure from physicians who believed that tobacco should be dispensed as a medicine and, therefore, kept for their own profit. In the same year, James increased the tax on tobacco by 4,000 per cent so that, if he failed to suppress it, he could at least profit from what was known (as early as 1610) to be a habit after Francis Bacon noted that it was hard to abstain from tobacco once its use had become customary.

In Russia, the Romanovs were more determined and maintained a ban on tobacco, on pain of death, from 1613 to 1689. This followed a ban imposed in 1612 in China. There are suggestions that countries that imposed bans were more concerned about the effect on their balance of payments resulting from the spread of an imported product than with the health of their citizens.

In spite of these attempts, smoking as a habit spread rapidly. Even in the 16th century, smoking was so commonplace in Spanish colonies that priests would smoke during Mass and while celebrating Communion. In 1575, the Roman Catholic church outlawed smoking in the Spanish colonies and, in 1624, Pope Urban VIII issued a Papal Bull against tobacco. In 1725, Pope Benedict XIII revoked the ban as dignitaries of the church would 'pop out' for a smoke during services.

Not all rulers were opposed to tobacco. Napoleon was a regular snuff taker and, in meetings, would signal one of his counsellors to hand over his snuff box, which was frequently pocketed by the emperor. Generally, they were returned later, though whether by Napoleon or by Josephine is unknown. Sometimes, a different snuff box was returned and courtiers took to having simple wooden or cardboard snuff boxes, knowing they would be taken and, possibly, replaced by a jewel-encrusted, gold box.

By the 19th century, many concerns were raised about smoking but its actions and effects were still not well understood. In *American Medicinal Plants*, published in the last quarter of the century, Charles F. Millspaugh (1887) believes the balance of the available information suggests that tobacco is not harmful in moderation but points out that the actual amount that constitutes 'moderation' will depend on the individual user. He is critical of much of the writing on the subject, stating:

> Concerning the many essays that are written upon this subject, the fact that all of them show to a careful reader whether the writer is a user or not, renders them very unsatisfactory and more or less faulty through partisanship.

He notes a long list of symptoms caused by tobacco from confusion to heart palpitations, exhaustion, spasmodic contractions and many others but suggests that it acts as its own antidote

WHAT IS THE BIGGEST KILLER? PART 2

since these symptoms subside with continued use. Furthermore, he notes their return if use is discontinued as proof that the drug is its own antidote.

Today, many governments are once again attempting to reduce the use of tobacco whilst being heavily reliant on the taxes it generates, and there is much more research on its effects and the behaviour of its addicts. A 2007 survey of smoking in New York City showed that overall 17.5 per cent of the adult population were smokers. This was made up of 15 per cent for women and 20 per cent for men. These figures show reductions since 2002. The survey, which involved researchers telephoning over 18s, showed that 7 per cent of men and 5 per cent of women reported themselves to be 'heavy' smokers defined as over ten cigarettes a day.

Sixty-five per cent of respondents reported that they had tried to give up smoking in the previous year but only 17 per cent of those who tried were still non-smoking at the time of the survey. As the survey states, the questions about non-smoking only asked if the respondent had given up, not for how long, so the figures would include people who gave up very recently and are likely to restart. The net level of long-term 'quitters' is likely to be much lower.

One news report based on the survey compared sales of nicotine therapy products in New York with the survey results to conclude that only a minority of people seek assistance when trying to give up smoking. It did not consider the possibility of the over reporting of attempts to quit since many smokers feel they should be trying even if they don't actually stop or reduce smoking at all. The survey also found a relationship between heavy drinking and smoking, and psychological problems and smoking.

A paper from the Monell Chemical Senses Centre reports that nicotine in the breast milk of lactating mothers who smoke cigarettes disrupts their infants' sleep patterns. Researchers measured the feeding behaviour and sleep patterns of 15 breastfed infants on two separate days. The mothers were smokers who abstained from smoking for at least 12 hours each day. On the second day, they smoked from one to three cigarettes immediately before the study of the children's behaviour began. Samples of the milk were taken and the nicotine content established. Total sleep time over three and a half hours declined from an average of 84 minutes when mothers refrained from smoking to 53 minutes on the day they did smoke. The level of sleep disruption was directly related to the dose of nicotine.

According to a Dutch study people over the age of 55 who are smokers put themselves at a far greater risk of developing dementia than people who do not smoke. The study involved almost 7,000 people age 55 and older for an average of seven years and found that current smokers at the time of the study were 50 per cent more likely to develop dementia than people who had never smoked or had given up.

A report from the Royal College of Physicians (RCP) says that smoking is declining by only 0.4 per cent a year in the UK and calls for a change of approach to smokers. The report cites the experience of using harm reduction strategies in drug addiction. This approach recognises that the addict craves the experience of the substance but looks for less harmful ways of satisfying the craving either by using a different substance or by finding a safer method of delivery.

For smokers, Professor John Britton, chairman of the RCP Tobacco Advisory Group, says the way forward is to find alternatives that deliver the same dose of nicotine but do not involve smoking. The existing nicotine replacement therapies are intended to help with a complete cessation of smoking and, therefore, have only a low dose of nicotine to help to wean the smoker away from the addiction. Prof. Britton's paper calls for the development of nicotine products that would fully replace the addict's need.

In another new study, conducted by the University of Buffalo, researchers found that the progress of multiple sclerosis was faster in smokers and former smokers than in non-smokers. There was no discernible difference in the rate of progression of the disease between current smokers and those who had, at any time in the past, smoked for a minimum of six months.

WHAT IS THE BIGGEST KILLER? PART 2

Research, conducted at the University of Chicago Medical Centre and published in February 2008, indicates that the effects of nicotine and opiates on the brain's reward system are equally strong in a key pleasure-sensing area of the brain – the nucleus accumbens. The work suggests that, although nicotine and opiates, such as heroin, are very different drugs, the effect on dopamine production is almost identical. It is hoped that the work will help to find ways of overcoming addiction but it seems that addiction to smoking may be as strong as heroin addiction.

So, it seems that an addiction to tobacco, once established, is very difficult to overcome and the long-term answer to the problems caused by tobacco addiction may be to reduce the number of new addicts. What is somewhat depressing is that, based on the number of new smokers each year, 90,000 children a day smoke their first cigarette.

There are three truly depressing estimates about smoking with which to end this chapter. Based on current projections for the progress of addiction to tobacco, half of the 90,000 children who first smoked today will have their lives shortened by smoking related diseases. The current estimated 5 million deaths a year will rise, within the next decade or so, to 7 million shortened lives every year. Overall, one billion people will lose their lives to tobacco during the 21st century.

6

Have You Got Something Undetectable?

Occasionally the acronym FAQ is out of place. Some questions were FTQ – Frequently Thought – but not actually asked out loud. Talking to a group of visitors to the Poison Garden I would say, 'One question many people ask is "Have you got something undetectable? I want to be able to get away with murder".' Quite often, at that point, I would see couples look at each other and smile because they had, indeed, asked that question between themselves on their way to the garden but hadn't expected it to be asked out loud.

My answer to the question was always the same – the way to get away with murder is to not have anyone suspect murder has taken place. If there is no investigation of a death because it is not believed to be the result of murder, the murderer cannot be discovered.

In terms of using plant poisons as murder weapons there is one essential tool: a time machine. The ability to go back to a time before forensic science had reached the point where almost all substances can be detected either as themselves or via the compounds they produce in the body, the metabolites, would give a much greater chance of avoiding discovery than exists today.

In this chapter we'll look at some of the plants that could have been used as murder weapons when their presence might have been confused with some innocent cause of death. We will also look at some of the difficulties for investigators in obtaining evidence to convict murderers in the absence of the scientific knowledge that we take for granted in the modern world.

Naked ladies – deadly *femmes fatales*

Colchicum autumnale, known as meadow saffron, autumn crocus and naked ladies amongst others, contains colchicine and colchiceine, the former being the more toxic. Colchicine destroys white blood cells and is used medicinally where excess white cells are a problem but the dose of colchicine that may be fatal is relatively small. Its common name comes from its unusual growing habit where the leaves appear in the spring, die off in the summer and the flower appears, directly from the ground, in the autumn. The appearance of the leaves with no flower has led the plant to be confused with wild garlic with fatal consequences.

In 2003, a 76-year-old man with a history of alcoholism ate the plant in mistake for wild garlic. He suffered renal and liver failure and died from cardiovascular collapse and respiratory failure. Previously, also in Central Europe, two people were poisoned by eating Colchicum autumnale instead of wild garlic. One died after 48 hours of heart, kidney, liver and lung failure whereas the other recovered after three days of severe gastrointestinal upset. In Japan, a 48-year-old man died, in 2002, four days after eating the plant. In another case in Japan, an 80-year-old woman died nine days after eating the bulbs of the 'Chinese lantern lily'. Doctors were unable to reverse the decline in her white blood cell count. Colchicine was found to be the cause, this being the first time the alkaloid had been identified in the Sandersonia aurantiaca.

In another case from Slovenia, a 71-year-old woman survived after mistaking Colchicum for wild garlic but only after exhibiting new symptoms up to three weeks after ingestion when her hair fell out. Another victim reported episodes of hair loss up to three years after ingestion.

HAVE YOU GOT SOMETHING UNDETECTABLE?

Dr Harvey Wickes Felter, MD (1922) offers this excellent description of the symptoms of colchicine poisoning:

> Upon the skin and mucosa colchicum is irritant, causing smarting and redness, sneezing and conjunctival hyperemia. Small doses increase the secretions of the skin, kidneys, liver, and bowels. Large doses are dangerous, producing gastric discomfort, nausea and vomiting and purging, and violent peristalsis with much intestinal gurgling. Poisonous doses produce a violent gastro-enteric irritation, with symptoms much like those of cholera, agonizing griping, painful muscular cramps in the legs and feet, large but not bloody evacuations of heavy mucus and serum, thready pulse, collapse, and death. Toxic doses are almost sure to kill in spite of efforts to save life, the patient dying a slow, painful, and agonizing death, the final act of which is respiratory paralysis. Consciousness remains to the end.

Colchicine can be used to treat gout, but it is important to keep the dose small enough to avoid its purgative effect. Until recently it was available in the USA in a variety of 'herbal' medications, that is, preparations which were not subject to the same scrutiny as prescribed pharmaceuticals. The US Food and Drug Administration (FDA) has, however, ordered companies to stop marketing unapproved drug products that contain colchicine in an injectable dosage form. Colchicine has been injected to treat acute gout attacks but the FDA says the therapeutic index of the substance is so small that dangerous overdose can easily occur. It is believed to have caused 23 deaths from use in this way though no time period for those deaths is stated. Three of the deaths occurred in March and April 2007 as a result of an error in a pharmacy, which resulted in a batch of colchicine eight times normal strength being sold.

So, the ability of colchicine to kill is undoubted and it might make a useful murder weapon if the victim is a known gout sufferer and, thus, might be believed to have taken an accidental overdose. Alternatively, the prospective murderer would be well advised to find a handy cholera outbreak before poisoning his victim since the likeness of the symptoms of cholera and colchicine is such that further enquiry would be unlikely.

Dependable Datura?

The Datura/Brugmansia genera have already been mentioned in the context of accidental death from experimenting with the psychoactive properties but Datura is also thought to have been widely used as a murder weapon. The variety of symptoms found in actual cases of accidental poisoning would seem to be a disincentive to this use, since a murderer wants to know what his victim is likely to do, yet its production of unconsciousness seems to be the property it was best known for in the past.

In medieval Europe it was associated with witches and witchcraft, after being brought to Europe by Gypsies. On the Indian subcontinent and Russia it was known as 'knockout drops', which thieves and prostitutes used to knock out their victims. Its effects are described by Carlos Castaneda in *The Teachings of Don Juan*. Datura has a widespread reputation for use during the commission of crime, though all the references are to second-hand reports rather than details of specific incidents. In the Middle Ages, especially in Italy, professional poisoners would, it is said, concoct a brew of Datura that is claimed to have been almost painlessly fatal. That alleged quality of deadening the senses before death, or during the perpetration of a crime, made it of the greatest value to criminals.

So well known was this ability that Christoval Acosta, a Portuguese naturalist and physician, wrote, in 1578, that Hindu whores gave it to their patrons because:

> these mundane ladies are such mistresses and adepts in the use of the seed that they gave it in doses corresponding to

as many hours as they wish their poor victims to be unconscious or transported.

The difference between the alleged effects of Datura, from a time before close study of the cause of death, and its known effects today may suggest that it was not as widely used as the old texts would have us believe.

An examination of 29 accounts of Datura poisoning, both scientific papers reporting on specific cases and 'herbals' going back to Dioscorides, gives a very variable view of the effects of the poison. In Chapter 4, we saw that Datura poisoning has been known to produce either sleep or agitation and convulsions with death resulting in about one third of instances. Dilation of the pupils seems to be the only constant symptom, which, of course, is not helpful if you hope to avoid the suspicion of poison being involved in the death.

On only a few occasions is mention made of the muscle weakness, which was supposed to make Datura a useful murder weapon by rendering the victim helpless; and memory loss, supposed to help whores get away with robbing their clients, is also only given in a minority of the sources.

The difference between what people thought and wrote in the past and today's attitudes is very well illustrated in a book published in 1887. In *American Medicinal Plants*, Charles F. Millspaugh, in discussing whether the Datura stramonium, thorn apple, is a native to North America or an import from Europe, remarks that it is only found close to habitations and, in one of those instances of casual racism, which seem so outrageous to us today, notes that the 'American Aborigines' call it the 'White Man's Plant' because it is only found 'near the homes of the civilized'. Though tending to the view that it is not native in the USA, Millspaugh cites a number of ways in which Native American tribes used it in their rituals.

Millspaugh also says that Baron Storck was the first to use stramonium medicinally. In the custom of the time he used it to treat diseases whose symptoms matched those caused by the plant. Thus it was used to treat mania, epilepsy and as a narcotic. It was believed to cause nymphomania and was, therefore, used to treat this 'condition'.

Marathon runners, mole catchers and multiple murder

The third favourite murder weapon relies, like the Colchicum, on being mistaken for an entirely innocent cause of death.

Strychnos nux-vomica is known as the poison nut tree because the nut-like seed contains strychnine, an indole which is 'a powerful central nervous system stimulant that competes with the inhibitory neurotransmitter glycine, producing an excitatory state with hyperreflexia, severe muscle spasm, and convulsions' (Palatnick, Meatherall, Sitar and Tenenbein, 1997). Put simply, the muscles contract uncontrollably. Death is usually the result of exhaustion or cardiac failure and post-mortem examination sometimes shows that the powerful contractions have caused muscle to be torn away from bone, which is why some victims contort into impossible positions.

Prior to the extraction of strychnine, nux-vomica was widely used for a variety of medical conditions. The name, however, does not, as many people believe, mean 'no vomiting' or 'stops vomiting'. 'Nux' is Latin for 'nut' and 'vomica' means 'lump' or 'abcess'. In the context of this plant, then, nux-vomica means 'the seed looks like a nut and has a lump on it'. Its use as a stomach medicine by those who mistranslate the name is not justified though homeopathists seem to suggest its use for just about every condition known to man.

In small amounts, strychnine acts as a stimulant and has, in the past, been used as the antidote to poisons such as hemlock, which depress the nervous system. At the 1904 Olympics in St Louis, the first man to cross the line in the marathon, Fred Lorz, had been seen waving from the backseat of a car for much of the course.

HAVE YOU GOT SOMETHING UNDETECTABLE?

The gold medal was, therefore, awarded to Thomas Hicks, a British born man who ran for the US. Hicks had been on the point of collapse ten miles from the finish so his trainers gave him strychnine sulphate mixed in egg whites and washed down with brandy. One dose was not enough and Hicks crossed the line being held up by two trainers. His legs were running but his feet were not in contact with the track. He required four hours of medical treatment before he was able to leave the stadium.

This was just one of many failings in the running of the 1904 games, which led the International Olympic Committee to hold another games, in 1906 in Athens, to try and restore the reputation of the event. It was said, at the time, that an event in Greece would be held every four years but the planned 1910 event was cancelled on the grounds of political uncertainty in Greece. The 1906 games results are not included in the official records of Olympic events.

It is sometimes said that strychnine builds up in the body over time in the same way as arsenic. This is not true. Strychnine is metabolised quite quickly. The body does, however, become sensitised to strychnine so that repeated exposure to small doses increases the effect. In time the same dose has a much greater effect, which may explain the belief that it remains in the body until a fatal accumulation is achieved.

Strychnine was first isolated from the seed in 1818 and, it appears, soon became a useful murder weapon. It was found to be an effective rat poison and, in the rat-infested inner cities of the time, most shops sold rat poison and anyone could purchase it without question. Strychnine has a very bitter taste, which could be problematic. If one's intended victim notices an unusual taste and comments on it, even if they proceed to ingest a fatal amount, their comments could cause enquiries to be made. So, a certain amount of culinary skill was required in order to disguise the taste either by adding something sweet to balance it or by adding it to something expected to have a bitter taste so as to mask it.

As already mentioned, strychnine is a muscle stimulant, which in a fatal dose may cause death by heart failure or by exhaustion from the uncontrolled contractions of other muscles. Typically, a strychnine victim lying on their back will arch up so that only the head and feet are in contact with the bed or other surface. This condition is called opisthotonos. These convulsions and contractions are described as tetanic, meaning they look like the symptoms of tetanus, which was endemic in the hovels of the time.

Thus a 19th-century would-be murderer could easily obtain strychnine; using their culinary skill they could then feed it to their unsuspecting spouse and when he died, the death could be attributed to tetanus. At this distance, it is impossible to say whether there were unnumbered murders committed in this way or whether this *modus operandi* features more in modern fiction than it did in fact.

The first suspected strychnine murderer was Thomas Wainewright (1794–1847) who was believed to have murdered three people in order to claim the insurance he had taken out on them but, due to the lack of reliable scientific evidence, he was only tried and transported for fraud. Wainewright's possible crimes occurred in 1828, '29 and '30 only ten to twelve years after the discovery of strychnine when little was known about it. At the time, the deaths were assumed to be due to tetanus. Wainewright fled to France and was not arrested until 1837 so murder charges could not be sustained and he was tried in connection with the insurance payouts, which resulted from the deaths.

The first successful prosecution for murder using strychnine came in 1856 when William Palmer was convicted of the murder of John Parsons Cook. The scientific evidence at the trial was that no trace of strychnine had been found in the corpse but the circumstantial evidence was strong. Palmer, a doctor, had obtained strychnine and, while attending the post-mortem as Cook's physician, had successfully contaminated the stomach and its contents by apparent carelessness.

What undoubtedly secured Palmer's conviction, however, was

the succession of expert witnesses who presented their experience of tetanus and pointed out the differences between any sort of death from tetanus and Cook's dying moments. Though only tried for this one case, Palmer had lost four children in quick succession some years before and both his wife and mother-in-law died in suspicious circumstances. There is reason to believe he may have been involved in a total of twelve deaths by strychnine poisoning. Had he been acquitted of Cook's murder, the Crown was ready to charge him with killing his wife since the inquest on her death, held before his trial for Cook's murder, had returned a verdict of murder. Palmer was, almost certainly, an early example of the medically qualified serial killer.

Palmer brought great notoriety to his hometown of Rugeley in Staffordshire and a law was passed to enable him to be tried in London where the chance of finding an unbiased jury was better. When he was hanged, it is said that the people of Rugeley petitioned the government to change the name of the town. The Prime Minister said he would agree but only if they adopted his name for their town. Apparently, the people of Rugeley didn't take too long in deciding not to accept Lord Palmerston's offer.

By the late 19th century, concerns were widespread about the dangers of the free availability of strychnine and it became more tightly controlled so that it was mostly used by doctors. Dr Thomas Neill Cream, a physician with a homicidal hatred of women, was convicted of four murders of prostitutes in 1891. He convinced them he had a pill that would prevent them catching diseases. It was strychnine.

Cream had an interesting history. He had been born in Scotland but moved to the USA with his family when five years old. He qualified as a doctor but was found guilty of the 1881 murder of a patient, Daniel Stott, presumably because he was having an affair with the man's wife. He might have got away with it, because Stott was epileptic and the death was attributed to natural causes, had not an anonymous letter been sent to the authorities claiming that death was due to a mistake by the pharmacist who supplied the victim. The letter had been written by Cream because he did not want Stott's death to go unremarked.

Amazingly, he escaped the death penalty and was released in 1892. He immediately moved to London and began his murderous pursuits. Again, it was his desire for notoriety that brought his downfall as he wrote letters to perfectly innocent people claiming he knew them to be responsible for the deaths of one or other of Cream's victims and attempting to blackmail them. He also, anonymously, offered his services to the police, portraying himself as a private detective.

Even at the moment of his death, on 15th November 1892, he couldn't resist one final attempt to ensure his notoriety. As the trap was about to drop he called out 'I am Jack'. His wish may have come true because, even today, there are those who say he was Jack the Ripper even though he was still in America when Jack's killings started.

In the 1940s, Dr Marcel Petiot was convicted of 24 murders committed during the Nazi occupation of Paris. His victims were mostly wealthy Jewish people whom he claimed to be able to help escape from the Germans. The victims were told to come to a house, owned by Petiot but not occupied, bringing all their wealth and valuables since the escape route required much bribery. Petiot had created a small room in the house with a false door said to lead to the outside. It is believed he offered his victims some last refreshment before their hazardous journey, laced with strychnine, and then ushered them into this room telling them their escort waited outside the false door. At his trial, Petiot claimed he had only killed collaborators and insisted the true number of his victims was 63 and not the 24 where the prosecution was able to collect sufficient evidence for a certain conviction.

But not all strychnine murders were the work of doctors. In 1931, Lieutenant Hubert George Chevis died after eating poisoned partridge. He is reported to have eaten only one mouthful and become ill within minutes before dying the next day. His wife was taken ill but recovered. Strychnine was identified as the cause of death. Confirmation that he had been murdered came on the day of his funeral when his father received a telegram from Ireland which simply said 'Hooray, hooray, hooray'.

In 1934, Ethel Major used strychnine, which her gamekeeper father had for mole control, to murder her husband. His death was attributed to 'status epilepticus', a condition producing prolonged seizure with convulsions, and she might have got away with it had she not put some leftover food scraps out in the garden where a neighbour's dog ate them and died as a result. An anonymous letter told the police about this and Arthur Major's funeral, which Ethel Major had been anxious to have as soon as possible, was postponed so that post-mortem examinations could be performed on Arthur and the dog. These provided evidence of the presence of strychnine and, when a search of her home found a key to her father's poison store, which he said had been lost some time before, a jury had no doubt of her guilt.

Use of strychnine-based poisons by mole catchers continued until very recently. In February, 2008, the UK government lost its appeal against having to enforce new EU legislation on the use of strychnine-based poisons to kill rats. The UK government was concerned about the 3,000 licensed mole catchers who, until then, could use strychnine in their work.

The most recently reported case of strychnine poisoning was chronicled in the October 2002 edition of the journal *Critical Care* which carried a paper about a 42-year-old man who took himself to hospital after swallowing some white powder he found in his shed where he had been drinking a whole bottle of wine. Shortly after arrival, he suffered cardiac arrest and was resuscitated. He suffered muscle spasms so severe that he was given a paralysing agent and his blood pressure remained very low for some time. The twitching continued for three days but, after five days, he returned to normal and apparently suffered no lasting damage. The paper notes that he remained in hospital a further five days for psychiatric assessment. Though unsaid, it may be that swallowing the powder was intentional.

IS THAT CAT DEAD?

Inheritance, ill-health and ill-gotten gains

The three plants discussed in this chapter have been responsible for an unknowable number of deaths and, there are others, such as Atropa belladonna, deadly nightshade, and Papaver somniferum, which have also found use for murderous purposes but, historically, the largest number of murders has been due to a metallic poison and not a plant.

It is sometimes said that Venetian ladies used Strychnos nux-vomica as the 'inheritance plant'. There is no evidence for this use. There was an 'inheritance powder', which has been shown to be arsenic, used as a way of speeding the acquisition of the family silver. Madame Toffana, who, in the 17th century, was accused of being responsible for 600 deaths mostly undertaken by women she had trained, used a paste containing arsenic. This was sold as a cosmetic but women who had the benefit of Madame Toffana's training knew how to apply it to more deadly purposes.

Arsenic was very readily available, is a cumulative poison enabling small undetectable doses to be administered over time, and would begin to produce symptoms long before death so that the victim would be known for their poor health prior to death.

Had the question been put 'What would you suggest as a good murder weapon?' my answer would have been 'None of the plants'.

7

But Aren't These Used as Medicines?

These days, perhaps because of a cynicism about the very large pharmaceutical companies, perhaps because of a sense of disappointment with the UK's National Health Service, many people are reverting to 'old-fashioned herbal remedies'. Personally, I'm with Dara O'Briain, the Irish comedian, on this one. O'Briain says the 'old-fashioned herbal remedies' that worked are, today, called medicines and everything else is just vegetable soup.

Certainly, some of the plants in the Alnwick Garden Poison Garden do contain substances used in mainstream medicine and others are available as 'herbal' or 'natural' remedies. Still others are used in homeopathy. In this chapter we'll look at some examples of each of these three 'medicinal' areas.

Belladonna – Matters of the Heart

First, the medicines. In *A Treatise on Belladonna*, issued by Lloyd Brothers, Cincinnati, Ohio in 1905, the first reference to the extract of Atropa belladonna being used medicinally is said to be in a book published in Paris in 1504 though there are earlier references that could be to this plant. However, in 1598, John Gerard in his *Great Herbal* is not convinced of the wisdom of keeping it around.

He says the deadly nightshade causes sleep, troubles the mind and brings madness if a few berries are eaten but more will bring death. The green leaves have been recommended as a medicine but Gerard (1598) suggests that it is best to:

> Banish it from your gardens and the use of it also, being a plant so furious and deadly: for it bringeth such as have eaten thereof into a dead sleep wherein many have died.

In evidence he cites the case of three boys of Wisbech who 'did eat of the pleasant and beautiful fruit hereof, two whereof died in less than eight hours after that they had eaten of them'. The third was saved by being given a mixture of honey and water causing him to vomit, often.

Gerard says these 'pernicious plants' should be removed from the garden and all areas where children and women may find them since women and children 'do oftentimes long and lust after things most vile and filthy' and, so, will be very inclined towards a 'berry of a bright shining black colour and of such great beauty'.

The potential harm of Atropa belladonna resulting from the attractiveness of the berries is further enhanced by its taste. We've already seen how both Senecio jacobaea, ragwort, and Aconitum napellus, monkshood, are not nearly as harmful as they are poisonous because their unpleasant taste deters ingestion. But the berries of the Atropa belladonna are not unpleasant. I say 'not unpleasant' because I don't agree with those who say they 'taste really good'. It was Mrs Grieve in *A Modern Herbal* (1931) who described them as 'intensely sweet' and many other references seem to have accepted this definition. In *A Treatise on Belladonna*, Felter (1922) says they have an 'insipid, but not unpleasant, taste'. Others have described it as 'sweetish' and my own experience of tasting a berry agrees with these descriptions. The berries are also very juicy, so an attractive berry, full of juice and not unpleasant to taste but potentially fatal, might be expected to have caused many deaths over history.

That it has not may be attributable to the folklore associated with the plant. It is said to belong to, or be cared for by, the devil and anyone else touching it will cause the devil to appear, wanting to know why you are stealing his plant. It seems our ancestors knew that if you said to a child, 'Don't eat those berries you'll get sick', they would give one to a friend or sibling for them to try so they could see how sick. I said this once, during a tour of the Poison Garden, and a female visitor piped up, 'That's what my brother did to me.' Apparently, when she was in her pram her brother gave her a deadly nightshade berry to see what it did to her before deciding whether to have any himself. What it did to her was put her in hospital for three days.

One of the best known effects of Atropa belladonna is to cause the pupils of the eye to dilate, a condition called mydriasis. Venetian women knew that having dilated pupils would make them more attractive to men so would put juice from the plant into their eyes before appearing in public. These are the same women who would give their husbands Madame Toffana's arsenic-based 'inheritance powder' so their attempts to snare a man may not have been entirely romantically driven. Telling that story, one day, showed me how far some people are willing to go to think the best of people.

Having the pupils dilated more than is necessary for the available light is an uncomfortable and distressing experience. A pharmacist told me of the time he was working with atropine, the alkaloid that gives Atropa belladonna its property of causing what is called mydriasis, and didn't realise a little had splashed into his left eye. He said his drive home, through the dark winter rush hour in Newcastle-upon-Tyne was frightening and, when he realised why all the lights were so bright, he gave up his journey until the eye had returned to normal. I would imagine there would be some correlation between the lights of moving vehicles and the flickering candle light of a grand Venetian ball so these ladies were prepared to suffer to snare a rich catch.

Mydriasis has been shown to occur when someone is sexually aroused. Tests have also shown that people respond more favourably to those who have dilated pupils so, it would seem, these Venetian ladies were using a technique based on science.

One visitor was not convinced that the whole process – from belladonna in the eyes to arsenic in the husband's food – was pre-planned. He felt that, being unable to see clearly in the excessive glare, these poor women may have realised too late that they had pledged themselves to an ugly beast rather than a handsome Adonis and murder was their only way to clear the slate and start again.

Though belladonna was in use medicinally since at least the 15th century, John Parkinson, in his 1629 herbal *Theatrum Botanicum*, has another idea of its benefits. He recommends an infusion of nightshade in wine as a 'good jest for a bold unwelcome guest' as it will ensure that the guest 'shall not be able to eate any meate for that meale nor untill he drinks some vinegar'.

It was the property of dilating pupils that led to the extraction of atropine and its use in two ways. In the 19th century there was much interest in the narcotic properties of a substance called esere, which was an extract from the Calabar bean. In the 1850s, a Dr Thomas Fraser, from Edinburgh, noticed that one side effect of esere was to cause contraction of the pupils. This led, directly, to the discovery that esere could be used as an antidote to atropine and vice versa.

Atropine is also used by ophthalmologists who need to see into the back of the eye and is still helpful for this purpose today. It can also be used to treat a condition where the muscles that move the pupils do not operate correctly and the vision is impaired as not enough light reaches the back of the eye. One visitor, who worked in an eye hospital, talked of a case where a young girl was prescribed atropine drops, which were to be kept in the refrigerator. Neither she nor her mother realised that the top was not securely on the bottle, which had fallen onto its side, until the mother began hallucinating after making herself a ham sandwich using ham that had been on a plate on the shelf under the bottle of atropine.

BUT AREN'T THESE USED AS MEDICINES?

And that is one problem with the use of atropine: its ability to cross what is called the blood/brain barrier and produce psychoactive effects. Another property of atropine is that it stimulates the heart and this property was used by anaesthetists to speed the revival of patients at the conclusion of an operation under general anaesthetic. Unfortunately, some patients, especially the elderly according to the 'gasman' who told me this, were waking up suffering mild hallucinations and becoming very confused about what was happening to them. There is now an artificial alternative that stimulates the heart but does not enter the brain.

Not all use of atropine as a heart stimulant has come to an end, however. Over the three and a half years I was Poison Garden Warden at Alnwick, I met a number of soldiers who said they had been given syringes of atropine when on duty in Iraq and Afghanistan and told to inject themselves with it if they believed they were becoming lethargic after a nerve gas attack. The hope seems to be that the atropine would keep the heart going until proper medical attention could be given.

So, Atropa belladonna is a useful tool for mainstream medicine but its fatal properties are not limited to John Gerard's boys in Wisbech. There have been numerous cases of accidental poisoning over the centuries and one woman made use of atropine as a murder weapon. Marie Jeanneret, a Swiss nurse was convicted in 1868 of the murder of seven patients.

Jeanneret had an obsessive preoccupation with poisons, especially Atropa belladonna, and experimented on herself with the effects of atropine. She was unsuccessful in extracting usable material direct from the plant, at one point burning herself badly when a pot exploded, and feigned an eye condition in order to get atropine prescribed as medication. Over time, like Mithridates before her, she built up a tolerance to the effects. Unable to pursue further experiments on herself, she resorted to using patients as unknowing subjects of her experiments. She specialised in nursing people in their own homes, which gave her long periods alone with them. She was not suspected of involvement in their deaths because she was described as a loving, caring nurse who was most distraught when her charge died. With hindsight, it can be seen that Jeanneret was distressed because she would have to go and find another guinea pig.

Her crimes were discovered thanks to the arrogance of a doctor who refused to believe that his patient had not recovered from the treatment being given and had Jeanneret searched, finding an empty phial of atropine in her pocket. Her conduct and her behaviour in court led to long discussions as to her sanity and, eventually, she was spared the death penalty and imprisoned at St Antoine, Geneva, where she died in 1884. Her case is credited with having led to the abolition of the death penalty in the Canton of Geneva.

Opium – a life saver

Papaver somniferum, the opium poppy, gives us two of the most useful medicines known, morphine and diamorphine. In Chapter 4, we looked at the number of deaths due to the over use of diamorphine when used under its common name of heroin. But, used medicinally, morphine has contributed to untold numbers of lives being saved by relieving the pain of people severely injured in war and accidents and the suffering of those undergoing treatment for cancer of all sorts. In Chapter 11 we'll look at this plant in much more detail.

Digitalis and the Doctrine of Signatures

Plants of the Digitalis genus provide another widely used and life-saving medication. The foxgloves contain cardiac glycosides called digitoxin, digitalin, digitonin, digitalosmin, gitoxin and gitalonin, which during digestion produce aglycones that directly affect the heart muscles. By slowing the heart rate and making the contractions stronger digoxin, which is the normal pharmaceutical used, is a great help for patients with irregular heart rhythms or whose capillaries are leaky, meaning the blood is not doing its job of delivering oxygen to the cells. But this use of foxglove is comparatively recent. John Gerard (1598) says 'They are of no use, neither have they any place amongst medicines, according to the Ancients.' That said, he goes on to say that foxgloves will clear the chest of phlegm.

In 1775, Dr William Withering was asked to comment on a family recipe for the treatment of dropsy, which had come from an old woman in a village in Shropshire. Dropsy, a condition where the soft tissue swells due to an increase in fluid retention, was at the time treated symptomatically. That is, diuretics were used to remove the fluid. It is now known that congestive heart failure results in a build up of fluid in the lungs as well as the soft tissue. Withering's early experiments with foxglove were performed on poor patients who attended a weekly two-hour free clinic that he offered, and, in seeking to use the plant as a diuretic, he, by his own admission, achieved little success.

He was inclined to give up his work with the foxglove when he heard from his friend, a Dr Ash, that the principal of Brazen Nose College, Oxford had been cured of hydrops pectoris, a sort of dropsy of the lungs, by means of the root of the foxglove. When Withering was able to obtain a supply of dried leaves, giving him the chance to measure dosage more accurately, he embarked on a series of trials all of which he set down in detail, even those which failed.

By the time his *An Account of the Foxglove* was published (1785), Withering had demonstrated the benefit of using foxgloves to treat dropsy even though he assumed its success to be based on its properties as a diuretic rather than having a direct cardiac effect. As one of his case reports notes, however, he was aware of the slowing of the pulse caused by the use of Digitalis and the appalling consequences of overdose:

CASE CVI

November 2d. [1782] Mr S————, of B————h————, Æt. 61.

Hydrothorax and swelled legs. Squills [a diuretic] were given for a week in very full doses, and other modes of relief attempted; but his breathing became so bad, his countenance so livid, his pulse so feeble, and his extremities so cold, that I was apprehensive upon my second visit that he had not twenty-four hours to live. In this situation I gave him the Infusum Digitalis stronger than usual, viz. two drams to eight ounces. Finding himself relieved by this, he continued to take it, contrary to the directions given, after the diuretic effects had appeared.

The sickness which followed was truly alarming; it continued at intervals for many days, his pulse sunk down to forty in a minute, every object appeared green to his eyes, and between the exertions of reaching he lay in a state approaching to syncope [fainting]. The strongest cordials, volatiles and repeated blisters barely supported him. At length, however, he did begin to emerge out of the extreme danger into which his folly had plunged him; and by generous living and tonics, in about two months he came to enjoy a perfect state of health.

Withering's place in the history of the development of medicine relies on three things: his willingness to look at a folk remedy to see if it had any merit when most of his contemporaries would have scorned such an enquiry; his lack of vanity in publishing all of his trial results even those that indicated failures in his treatment of patients; but his greatest legacy is much more general than just the use of the foxglove.

At some point in his work, Withering had what may be described as an 'Emperor's New Clothes' moment. He realised that there was nothing about the look of Digitalis purpurea that would lead you to conclude that it could be used to treat dropsy and, against the then conventional wisdom, concluded that the Doctrine of

BUT AREN'T THESE USED AS MEDICINES?

Signatures, the idea that the look of a plant defined its medicinal use, was without merit. This revelation freed doctors from the bounds of only looking for remedies from plants that met the criteria of the Doctrine of Signatures and, some would say, began a much more rationally based investigation of disease and its treatment.

So extracts of foxglove, which are obtained today by growing Digitalis lantana commercially, are improving the lives of many patients and saving, at least, some of them. Overdose, however, leads to bradycardia, where the heart rate falls below 60 beats per minute and may lead to cardiac arrest. In 2005, an amateur botanist committed suicide by eating foxglove leaves. Crude plant material is strongly emetic and most people who eat it accidentally suffer nothing worse than vomiting but this individual knew to eat only a few leaves so as not to produce nausea.

And that is one of the downsides of foxglove and its extracts. It has what is called a low therapeutic index. The therapeutic index of a substance is the difference between the amount that does nothing, the amount that does good and the amount that does harm. There is greater risk with a substance with a low therapeutic index not only because a small overdose can be catastrophic but also because it offers the opportunity to disguise a deliberate action as not intending harm.

On 2 March 2006, Charles Cullen was sentenced to multiple terms of life imprisonment in New Jersey after confessing to 29 murders of patients at hospitals where he worked as a nurse. His preferred weapon was a lethal injection either of digoxin or of insulin. It is believed he may have killed another eleven but, it seems, their illnesses may have killed them before the drugs could have an effect.

Digitalis is another plant where some of the folklore seems to have been intended to keep children away from it. Foxglove is said to be a corruption of 'folk's love' meaning the fairy folk took advantage of its downward pointing flowers to obtain shelter from the rain or the sun. As home to the fairies, it was capable of bringing bad luck on a family if the fairies were disturbed so children were strongly advised to look but not touch. I have a friend who maintains that, on a warm summer's day, you can see the flowers vibrate and hear the fairies snoring inside the flowers. He is a beekeeper, though, which I suspect may have something to do with these effects.

As well as this preventive folklore, it has many other stories attached to it. 'Foxglove' is sometimes said to have come from the plant being given by the fairies to foxes to wear as gloves so as to leave no trace when raiding a hen house. This explanation was popular with people inclined to raid the rich man's hen house for themselves. It is sometimes suggested that the name comes from 'foxglew' an Anglo-Saxon musical instrument with a similar shape.

The plant has to be picked with the left hand, from the north side of the hedge for its potency to be retained and some sources add that it must be picked at full moon for greatest magical effect. When the tall stalk flowers it will nod in the direction of anyone with supernatural powers. Grown around the house, it protects against malign influences. Another story has it that God created the foxglove so its bell-like flowers would ring to warn foxes of hunters coming to cut off their bushy tails.

Vincent Van Gogh was treated by Dr Paul-Ferdinand Gachet for mania and/or, possibly, epilepsy. It is often said that Van Gogh's problems came from his addiction to drinking absinthe but there seems to be no definitive evidence for this. In the 19th century Digitalis was used as a sedative, an anticonvulsant, and an anti-manic agent and would have been an obvious choice for Dr Gachet's treatment. Van Gogh painted Doctor Gachet, twice, and in both pictures the doctor is holding a foxglove plant.

Regular treatment with Digitalis is known to cause xanthopsia, a condition which gives everything a yellow colour cast. Digitalis is also known to cause visions of halos of light. Both these phenomena are seen in Van Gogh's work, which suggests that he was having regular doses of Digitalis.

BUT AREN'T THESE USED AS MEDICINES?

Soup or superdrug?

So, Atropa belladonna, Papaver somniferum and Digitalis have all gone from herbal remedy to mainstream medicine but is Dara O'Briain right when he says the rest are just soup? I think the answer is 'almost completely'. Incidentally, by 'herbal remedy' I'm talking about a plant or its extract that is administered in a measurable amount. I'll get to homeopathy later. I have to say 'I think' because the truth is that very little solid research has been done on the effects of these remedies. That in itself says to me that they, probably, don't work because if they did there would be more interest in understanding how and exploiting them more fully.

But I said 'almost completely' because there is one remedy where we do know something of its action. Vitex agnus-castus is sold as 'Agnus Castus' and is recommended as a way of relieving the hot flushes associated with the menopause. It has been found that its active ingredients, volatile oils, stimulate the pituitary gland and increase the production of progesterone, one of the female hormones. It seems likely that regulation of the amount of this hormone may actually have an effect on the menopause. It also could explain how this plant gets it two best known common names: 'chaste tree' and 'monks' pepper'. It is said that, in the olden days, monks would eat the seeds of the plant in order to stop them thinking the thoughts that monks shouldn't think. Obviously, whilst additional progesterone might be helpful for a woman in the change of life, for a man its presence would be most likely to encourage him to pray and tend the monastic garden.

I should say that whilst the mechanism of action has been demonstrated it is not certain that the additional progesterone is reducing the severity of hot flushes. It remains entirely possible that there is some placebo effect going on. Which brings us to homeopathy.

Nothing works as well as homeopathy

Homeopathy is a very useful demonstration of the principle that it is not the substance that is the poison, it is the amount. Numerous visitors cited cases where they had taken a homeopathic remedy made from one of the most poisonous plants. Nux vomica from Strychnos nux-vomica is a particular favourite and is supposed to be capable of curing all manner of conditions but Aconitum, Atropa belladonna, Oleander, Datura, Bryonia, Actaea, Pulsatilla and Ruta all appear in homeopathy catalogues. No one ever got poisoned by taking a homeopathic extract of Aconitum diluted to the extent of one part in many, many millions.

And, of course, it's hard to say if anybody ever got cured either, at least, by the substance itself. I'm not saying that there is no possibility that a very minute amount of a chemical might trigger a curative reaction in the body, though it should be understood that at the higher dilutions favoured in homeopathy there is almost certainly not one molecule of the 'active ingredient' in a typical dose. I'm just saying this is not the basis of homeopathy. Homeopathy is all about triggering the placebo effect so that the brain believes a cure is being effected and produces that cure.

A simple example demonstrates this. If I take one litre of a substance and add ninety-nine litres of water, I have a one in one hundred dilution. If I take one litre of a substance and add nine litres of water I have a one in ten solution. If I now add ninety litres of water to the ten litres, I have a one in one hundred solution. Scientifically, there is absolutely no difference between the two solutions. Homeopathy, however, insists that the two are different because the repeated dilution, one plus nine in my example, but one plus ninety-nine in homeopathy, enables the water to 'retain the memory' of the active substance.

The effectiveness of homeopathy relies on the way it is administered. All too often, a rushed GP will tell you what he thinks you have and suggest a medicine that 'might' help. Most people have been told 'If you're no better in a few days come back and we'll try something else'. Homeopaths are not rushed and, after listening carefully to your symptoms, will tell you that they know what will cure you. Clearly, the placebo effect is much more likely to occur in the latter case.

Researchers are beginning to realise how powerful this effect is and work is being undertaken to try and understand it better. In a trial, one group was given a sugar tablet and told it was a cheap generic medicine while the other was told that their pill cost several pounds each. Those receiving the 'expensive' pill reported a statistically significant higher level of improvement in their condition.

In a normal blind controlled trial it was found that the control group, i.e. those who receive no medication but are followed to study the passage of the untreated condition, had a significant number of people reporting that their condition improved. In another trial, one group was given a placebo with minimal interaction with the medical staff while a second group, given the same placebo, had long consultations with staff who had been trained to talk about their good experience with the 'drug' being given and the people they knew who had benefited from it. By now, you should be able to guess which group had the higher reported improvement in the condition.

And, finally, medical ethics require that people are advised of potential side effects from any drug. This means that, in a double blind trial, the people who will be given the placebo get the same warnings as the people who actually get the drug. It has been found that a significant number of people on the placebo report, and in some cases exhibit, the side effects associated with the drug under trial.

In a trial in America using an extract from Cannabis sativa to control muscle weakness in multiple sclerosis, one of the trial subjects pulled out after reporting that the drug he was being given was making him 'high', a feeling he did not enjoy. He was receiving the placebo and was getting high because his brain thought that was what would happen.

Ah, but, say the supporters of homeopathy, it works on animals so it can't just be placebo effect. Oh, yes it can. As above, in one trial the control group reported feeling better just because someone was paying attention to their condition. Reports of animals getting better after homeopathy may be just the animals responding to increased human attention. The so-called 'observer's paradox' refers to the phenomenon where the mere presence of an observer at an event or experiment influences the outcome. Taking more care of the animal may be enough for its brain to begin to correct the problem it is having. Or, it may also be simply that the animals were about to recover spontaneously, which is often what happens with humans. Many viral infections are self-limiting and will go away with no treatment but people who have taken a homeopathic remedy assume that it has produced the 'cure'.

Few if any pet owners who use homeopathy on their pets only give the homeopathic remedy. They change the animal's diet, change its environment and offer extra love and attention but, when recovery occurs, they attribute it to the homeopathic remedy. But it doesn't really matter what worked, as long as the animal or human patient recovers, that is what is important. And that can, to some extent, be said for mainstream medicines and herbal remedies as much as for homeopathy.

In the next chapter we'll look at situations where the effect of the substance needs to be determined with as much accuracy as possible – because a murder conviction may depend upon it.

8

Have They Been Used as Murder Weapons?

As well as the thousands of tours I've taken around the Poison Garden, I've given a lot of longer talks both in the Alnwick Garden and to external groups. The original idea was to give people all the information that could not be included in a 20-minute tour of the garden itself. When I realised that such a talk would last about four or five hours, I decided to break out particular areas to talk about and I now have a repertoire of nearly ten different talks, which I offer to groups who want to hear me speak. There is one talk, which, more than all the others put together, is requested by group secretaries; sometimes, it must be said, a little shyly. That talk is entitled 'Medical Murderers'.

There is something about us, the human race, which makes us revel in the dark doings of some of our number and, I've found, the more unpleasant the details, the better. So, in this chapter we'll look, briefly, at some of these notorious murderers and the plants they used and how they used them. I stress *briefly* as many of these cases have been the subject of whole books in their own right and some of them continue to arouse controversy.

The topic is 'Medical Murderers' because, from the mid-19th century onwards, the people with the easiest access to the extracts of poisonous plants have been the medical profession, both doctors and nurses. Rather than go through them in chronological order we'll look at them plant by plant in ascending order of the number of known deaths. The perpetrators are those who combined sufficient knowledge of poisons to enable them to succeed for some time in avoiding the suspicion of murder with an arrogance that, even if murder were suspected, they were of superior intelligence and could escape detection.

Everyone knows Dr Crippen

Ask most people to name a doctor from history who committed murder and they will say Dr Crippen. Though Crippen only murdered one person, his wife, and plenty of spousal killings have occurred throughout history, the case caught the public imagination at the time and has continued to do so. It has recently been argued that Crippen was wrongly convicted. In my view, those arguments cannot be sustained but the continuing interest in Crippen is such that they were the subject of a TV programme.

Dr Crippen's weapon of choice was hyoscine, the principal alkaloid in Hyoscyamus niger, black henbane. There are many reports of its effects when taken either accidentally or for its hallucinogenic properties. In January 1737, Dr Patouillat who practised in Toucy, France, was called to nine members of the same family who had become unwell after eating soup. The father of the family had been away at the time and was, thus, able to tell the doctor that he had dug up what he thought to be parsnip roots the previous day. These turned out to be the roots of Hyoscyamus niger, black henbane.

The symptoms Dr Patouillat describes, in a letter to M. Geoffroy, Paris, which was passed on to Sir Hans Sloane, are typical of the effects of henbane.

> The madness of all these patients was so complete, and their agitations so violent, that in order to give one of them the antidote, I was forced to employ six strong men to hold him ... the patients could give no account of their ailments nor of the quality [quantity] of the poison they had taken ... [Various remedies were administered] ... The next day

HAVE THEY BEEN USED AS MURDER WEAPONS?

I visited the patients and found them in a quite different condition; for they had all recover'd the use of their reason, but remember'd nothing of what happen'd.

All this day, every object appear'd double to them, that is, upon looking at a man, a beast, or a tree, they saw two.

I return'd to see them the next day, and found that the symptoms were removed; but were succeeded by another altogether as surprising, to wit, all objects appear'd to them as red as scarlet.

(Patouillat, 1737)

In the same journal, but in 1751, Dr John Stedman, surgeon-major to the regiment of the Royal Grey Dragoons, reported the case of seven people, five men and two women, who, in 1748, made a broth with the leaves. Stedman says the plant was Hyoscyamus albus, white henbane, but, in a commentary following the letter, Mr William Watson gives several reasons why this cannot be and concludes that the plant was Hyoscyamus niger.

As in other cases, Dr Stedman reports the rapid onset of symptoms and their seriousness:

> The man, who pull'd these leaves in mistake for another plant, said, that from the nearest conjecture he could make, there might be from fifteen to twenty leaves, boil'd in about ten quarts of water. They did not eat one half of that quantity, and the poison began to discover itself with some of them in half an hour...
>
> I saw them about three hours after having eat of it; and then three of the men were become quite insensible, did not know their comrades, talk'd incoherently, and were in as high a delirium, as people in the rage of a fever. All of them had low irregular pulses, slaver'd, and frequently chang'd colour: their eyes look'd fiery, and they catch'd at whatever lay next them, calling out, that it was going to fall. They complain'd of their legs being powerless... Next day they had no other

complaint than people commonly have after great drinking; but afterwards ... some of them complained [of various symptoms] ... above a month after the accident.

(Stedman, 1751)

A 34-year-old woman, treated by a Dr White, experienced symptoms within ten minutes of drinking a tincture of Hyoscyamus, thinking it to be a laxative known as a black draught. She suffered a burning sensation in her limbs, which lost their power, giddiness, intense thirst and a purple rash especially around her neck and face, which became swollen. When Dr White first saw her, four hours after ingestion, she was almost insensible and unable to speak. Her tongue was swollen and dry and her pupils dilated. Three hours later, she could hardly see and could not move her limbs. Twenty-eight hours after the poisoning, her condition began to improve but it was six days before the use of her legs began to return. She had no memory of what had happened and suffered ongoing short-term memory loss.

But the best report of the effects of the plant comes from Gustav Schenk (1956) who, when a young man, while staying in a remote village studying plants, decided to experiment with the effects of black henbane. He roasted some seeds and inhaled the fumes given off. In *The Book of Poisons* he describes the effects. His description is incomplete as he says that one of the key effects is to obliterate memory whilst leaving the overall experience and the visions seen intact.

> The Henbane's first effect was a purely physical discomfort. My limbs lost their certainty, pains hammered in my head and I began to feel extremely giddy.

All this began within a quarter of an hour of first inhaling the fumes. His dilated pupils made his vision hazy and, when he tried to look in a mirror, the mirror swayed making it hard for him to keep his face in view.

HAVE THEY BEEN USED AS MURDER WEAPONS?

I lost all recollection of my previous actions. I strove to figure out what had wrought such a change in me ... my confusion, drunkenness or illness was making my head ache and causing ... all the physical misery I suffered during those minutes.

He now begins to see visions of a yellow disc which, though plain, seemed to be casting glances at him in a way which made him shudder.

... this vision caused me tremendous amusement ... I imagined that my arm was talking, or my foot, and I answered them ... I was seized by a fear that I was failing apart ... I experienced an intoxicating sensation of flying.

Henbane was one of the plants used by witches to make a salve, which they would rub into the skin to induce the feeling of flying. In Chapter 9 we'll see what John Gerard had to say about these potions.

Though Schenk's visions were of soaring high above the ground he never lost the sense of being seriously ill and the conflict between the constant urge to move and the lack of the strength to move gave him great discomfort. When the visions subsided they were replaced with nausea and pain plus a 'grey misery' filling his mind, which lasted for some time.

The seed heads look like a piece of jawbone complete with a row of teeth. This plant was, therefore, in accordance with the Doctrine of Signatures, used in dentistry. Quite possibly, the hallucinogenic, soporific effects of the plant would have made people forget the toothache. This use dates from ancient times but, in Anglo-Saxon England, it was specifically believed to kill the worms whose presence was said to be the cause of toothache.

Anglo-Saxon folklore talks of a great battle with a giant worm, which resulted in the worm being cut into nine pieces and the nine pieces becoming the nine 'flying venoms', which were believed to be the cause of all pain and illness. There are suggestions that travelling 'doctors' would visit towns offering their proven cure for toothache. This involved getting the patient to open their mouths over a bowl of henbane seeds in hot water, the stated aim being to get these fumes into the worms and kill them. In fact, of course, the patient would breathe in the fumes and experience a deadening of the whole area. But, to add credence to his claims, the charlatan would use sleight of hand to deposit small pieces of lute string into the bowl so that he could show the patient the dead worms, which had dropped from the teeth.

A 15th-century leech book gives details of the use of henbane to kill worms in the teeth causing toothache but, instead of putting the seeds in hot water, it suggests mixing the seeds with leek seeds and flour and putting the mixture onto a hot tile. A hollow stem, big enough to cover all the mixture, is used to channel the smoke into the mouth so that it reaches the worms. The entry finishes 'This is proved'. *Macer's Herbal*, another mediaeval medical textbook, suggests soaking the root in vinegar and holding the resulting liquid in the mouth for a long time.

By the early 20th century, hyoscine was used as a travel sickness remedy so it is, probably, just coincidence that Dr Crippen, whose work involved selling patent medicines (many would say 'quack' remedies) for use in dentistry, chose hyoscine as his weapon. There has been a great deal written about Crippen's possible motives but it seems that the simple answer is the right one. He had been having an affair with the young typist at work who was anxious to put the relationship onto a proper footing and who had, probably, already miscarried the doctor's child.

What made the case of public interest was that Crippen and his lover, Ethel Le Neve, tried to escape to Canada when Crippen realised his crime was about to be discovered. For the first time wireless was used to report their presence on a transatlantic ship, and a chase ensued, lasting several days and widely reported by the popular press, with the police officer in charge travelling on a faster vessel to reach Canada before Crippen. But the most stunning part of the story, for those of us used to forensic crime novels and TV series, is that during the excavation of the remains of Belle Elmore, as Mrs Crippen was known, police officers handled some of the body parts with their bare hands and, at one point, the whole basement where the remains were found was sluiced out with a strong disinfectant to try and reduce the awful smell.

Murder by monkshood

We go from Dr Crippen with his one murder, motivated by lust or love, to Dr George Henry Lamson who committed two murders but only the second is of interest here. Lamson was a not very successful GP who was, apparently, always short of money. In spite of this, he travelled to the USA in 1881. On his return, his money worries multiplied and he made plans to go to France. On the day of his departure he visited his 18-year-old brother-in-law, Percy John, a cripple, who lived in at a school, Blenheim House, in Wimbledon. Lamson took with him some Dundee cake and a new invention he had brought back from the USA. Whilst Lamson, Percy John and Mr Bedbrook, the headmaster, drank tea and ate the cake, Lamson showed off this wonderful new idea, a soluble capsule making it possible to swallow medicine without tasting it. In front of the headmaster, Lamson filled a capsule with sugar and gave it to Percy John to swallow. Within an hour, after Lamson had left to catch his train, the young man began profuse vomiting and was dead within a couple of hours. Percy John had a small inheritance, which, on his death, would pass to his sister, meaning, under Victorian law, it would become Lamson's property.

Murder was immediately suspected and Lamson was the obvious suspect. This suspicion multiplied when a chemist came forward to say he had sold Dr Lamson some aconitine, the most powerful of the alkaloids in Aconitum napellus. The problem was that there was, at that time, no chemical test for the presence of aconitine. Which brings us to the key feature of the substance. In Chapter 3 I explained that the unique taste of monkshood and its active ingredients made it unsuitable as a murder weapon and, perhaps, Lamson thought that would help him avoid conviction since Percy John had made no complaint about the taste of anything.

The forensic investigator on the case, Dr Thomas Stevenson, used this unique taste to good advantage. At this point, when taking tours round the Poison Garden, I would always warn visitors that I was about to tell a story that would make them go 'Yeuk' and, generally, in spite of the warning 'Yeuk' from all directions would greet the punch-line. Dr Stevenson had never dealt with a case of aconitine poisoning but he had read about it and its unique taste. So, he obtained a sample and tasted a tiny amount to know for himself the taste profile and then, to see if it was the murder weapon, he tasted the dead boy's vomit. Now, that was usually enough to elicit the 'Yeuk' reaction so I didn't need to go on to say that he also tasted the stomach contents and parts of the bowel and other internal organs.

Having convinced himself that the murder weapon was aconitine, and confirmed that he wasn't just tasting what he expected to taste by having an assistant repeat the tests and suffer many hours of stomach discomfort as a result, Stevenson proceeded to inject some mice with aconitine and others with various extracts from the victim and compared their demise.

The forensic evidence was crucial in obtaining a conviction because there was no clear evidence on how the poison had been delivered. And that question, just how was Percy John poisoned, is still discussed today. Some say it was in the Dundee cake but all three men ate the same cake and suggestions that Lamson had precut the cake and poisoned just the piece he would give Percy John fall down because the trial transcript says the headmaster cut the cake. In any event, the taste would have been recognisable. It has even been suggested that the aconitine was injected into a currant or raisin in the cake, which Percy John might be expected to swallow whole and, thus, not taste the poison but that seems very far-fetched.

Those who say the poison could not have been in the capsule because Lamson filled it with sugar in front of the headmaster miss two points: the fatal dose of aconitine is so small that it could have been present in the capsule whilst leaving plenty of room for the sugar and, crucially, the trial transcript contains the headmaster's testimony that Lamson shook up the capsule, and commented on so doing. It seems to me he wanted to mix the aconitine with the sugar to avoid any possibility of a different powder being evident as Percy John took the capsule to swallow.

After his conviction, friends in America tried to use their influence to have pressure put on the British government to spare his life. They argued that his addiction to morphine meant he could not be considered to have been in his right mind. It is interesting to note that the prosecution spent a lot of time establishing Lamson's financial state in order to establish motive, even calling to court tradesmen to whom Lamson owed a few pounds, but they did not bring any evidence of an addiction to morphine, which would be a very good reason for Lamson to be short of cash. Equally, the defence would, surely, have presented evidence that Lamson's morphine addiction meant he was not in sound mind when he committed the murder and should, therefore, be spared the attention of the hangman.

It is quite clear that Lamson murdered Percy John and, though unprovable, it is widely believed that a few years before, he had murdered his other brother-in-law and taken his inheritance. If he did, that gives a useful moral to the story: if your brother-in-law is a doctor, don't leave him money in your will. It may be too tempting.

The next three plants in our ascending order of number of victims have been dealt with in earlier chapters. With a body count in single figures, Atropa belladonna is next on the list and, in Chapter 7, we looked at the case of Marie Jeanneret, but her behaviour was undoubtedly due to mental illness so her case does not fit the pattern of arrogance and power. Also in Chapter 7, I mentioned Charles Cullen who brings the body count due to plants of the genus Digitalis into the tens. In Chapter 6 we looked at poisons that might be mistaken for some naturally occurring killer disease, which brought us to Strychnos nux-vomica where we looked at the case of Dr William Palmer who, probably, thought that so little was known about strychnine that his murders would be safely attributed to tetanus. The other medical users of strychnine, Thomas Neil Cream and Marcel Petiot, simply believed they were too clever to be caught. Petiot's 63 victims added to Cream's 4 and Palmer's possible 7 take the total for strychnine above 70 but that number pales into insignificance as soon as we turn to the plant occupying the top spot in our chart of infamy.

To sleep, perchance to die

Papaver somniferum, the opium poppy, is, as we have seen, the principal cause of the approximately 200,000 drug related deaths around the world each year. The use of morphine as a murder weapon cannot hope to approach those sorts of numbers but the difference, of course, is that the victims of murder by morphine, in the nearly 200 years since its discovery, had no choice about their involvement with it.

Morphine was first commercially available in 1817 and, in 1823, it claimed its first murder victims. Auguste Ballet was, it seems, not much concerned with family love and loyalty. He wished his brother, Hippolyte, dead so that he might inherit his share of the considerable family wealth and asked his physician, Dr Edme Castaing, to take care of the matter for him for 100,000 francs. There is little detail available as to how Castaing managed to get Hippolyte Ballet to ingest a fatal dose of morphine but the job was soon done. Auguste Ballet's downfall came about because of his hatred of his sister, which made him susceptible to Dr Castaing's suggestion that if he, Castaing, were named sole heir in Auguste's will there could be no chance of the sister inheriting. Signing the will was signing his own death warrant and Dr Castaing promptly proceeded to try and secure his inheritance. Unfortunately, for him, ingesting morphine (the hypodermic syringe would not be available until the 1850s) is an uncertain means of administering a fatal dose and it took three attempts to secure Auguste's death. Since the two were travelling at the time, the murder took place at an inn some distance from Paris and it was Castaing's hurried trip to Paris to obtain more morphine that was a key factor in his conviction, there being, at that time, no certain way of demonstrating the use of morphine scientifically.

Forensic techniques were still of limited value in 1893 when Dr Robert Buchanan was tried, in New York, for the 1892 murder of his second wife, Anna Sutherland. Dr Buchanan divorced his first wife and married Anna Sutherland who was a wealthy brothel keeper. Buchanan's friends were shocked at his behaviour but were, nonetheless, sympathetic when Sutherland died soon after their marriage. Her death was attributed to natural causes, there being no outward sign of poisoning, and Buchanan might have escaped detection had he not done two foolish things.

Mistake number one was to remarry his first wife (though the wedding was in secret it did not remain so), and mistake number two came two years before when he was heard discussing a case in which a medical student murdered his girlfriend with morphine and was suspected because her body showed the contraction of the pupils, which is characteristic of a high dose of morphine. Buchanan was heard to say that the young man had been foolish because all that was necessary was to put atropine, from Atropa belladonna, in the eyes immediately and the two would work against each other and leave the pupils unchanged.

When a journalist, who suspected foul play, heard of this incident he brought the matter to the attention of the police who were able to obtain permission to exhume Anna Sutherland's body. Morphine was found to be present and Buchanan was charged. In court, it was expected that his defence would rely on the absence of pinpointing of the pupils so the prosecution brought in a live cat, gave it a fatal injection of morphine plus atropine drops in its eyes, and passed the dead body round the jury for them to see that its pupils were normal in size. This 'Grisly Demonstration', as it was called in newspaper reports at the time, was enough to reinforce the forensic evidence and Buchanan was convicted and executed.

There is no way of knowing whether Jane Toppan intentionally followed Buchanan's lead when she used morphine and atropine in combination.

Toppan is another example of someone whose early fascination with poisons shaped their career. Toppan was born Honora Kelley in 1857 but took the name Toppan from the family that informally adopted her when she was twelve. She realised that the best way to pursue her interest in poison was to become a nurse so that is what she did. In June 1902, she was tried for three of the eleven murders which investigations had suggested she had committed. These were the three victims whose bodies had been exhumed. Toppan started out using morphine as her weapon but moved on to using morphine and atropine. It seems she was trying to see how close she could get her 'playthings' to death before bringing them back rather than following Buchanan's lead, but this mixture disguised the contraction of the pupils expected in morphine poisoning. Though she claimed that she was sane because she knew she was doing wrong (that was what made it so satisfying), a panel of doctors judged her to be insane and she was held in a secure mental hospital until her death in 1938 at the age of 81.

The State of Massachusetts paid Fred Bixby to defend her after receiving evidence of her lack of funds. This is one remaining mystery from the case. There was clear evidence that she had robbed many of her victims; cash had gone missing and a number of pawnbrokers' receipts were found in her accommodation for items stolen from them, but no explanation has ever been found for how the money was spent. The day after her committal, Bixby announced that she had told him that, starting from her days as a trainee nurse, she had murdered at least 31 people and had, frequently, climbed into bed and hugged them in order to feel the life leaving them.

Forensic science continued to develop over the half century following the Buchanan trial but it remained the case that for a murderer to be discovered, murder had to be suspected. Dr Robert Clements was able to murder his first two wives without attracting suspicion. With the third, the suspicion did not arise until after her funeral and the evidence was too weak to cause an investigation. Perhaps Dr Clements was unfortunate that one of the police officers who knew of the small suspicions about the third wife heard about the 1947 death of Clements' fourth wife before the funeral could be held.

Like the previous three, the fourth wife's death certificate showed she died of natural causes but, in a change from his previous practice, Dr Clements had a Dr James Houston sign the certificate instead of signing it himself. Signing a family member's death certificate was not illegal at the time though it was officially frowned on. The police issued instructions for the funeral to be postponed so that a post-mortem examination could be performed. As soon as Dr Clements was advised of the postponement he gave himself a fatal injection of morphine because he knew that the post-mortem would provide irrefutable evidence of his guilt.

There is a bizarre aside to the case. When he learnt that Mrs Clements' death had been due to murder by morphine, Dr Houston also committed suicide. Contemporaneous reports suggested Houston was a weak character who may have accepted Clements' stated reason for his wife's death and meekly signed the certificate but, of course, it is possible that his involvement was more knowing.

IS THAT CAT DEAD?

So, in 1893, Dr Buchanan assumed he could defeat such forensic science as there was available but, in 1947, Dr Robert Clements knew the science would certainly see him convicted. In *Cause of Death* (1980), Frank Smyth discusses the case of Dr Buchanan. Smyth quotes an unnamed American toxicologist who is of the opinion that modern analysis methods, which can detect as little as 1/5,000th of a grain of morphine, mean that murder by morphine is a thing of the past as 'A modern doctor who wished to use morphine for murderous purposes would have to be an expert – much more skilled and much more wily in his methods than Dr Buchanan.' When this was published, Dr Harold Shipman was eight years into his 25-year career as a serial killer.

Because of the length of his murderous career plus the cremation of many of his possible victims making forensic examination impossible, the exact number of Harold Shipman's crimes will never be known but the enquiry set up under Dame Janet Smith to look into the affair concluded that it was probable that 284 died because Shipman wanted them to. As far as can be told, since Shipman never cooperated with any enquiry, that was his only motive in the vast majority of cases: he had the power to end life and he liked to use that power. In some of his last cases, the ones that brought him to the attention of the police, theft and fraud were also involved but they have to be seen as subsidiary to the desire to show that he could control events.

Incidentally, his suicide was not some show of remorse. It was a further demonstration of his control of outcomes. Though Shipman's pension rights were forfeit upon conviction, his wife's widow's pension was not. By dying before his sixtieth birthday, Shipman secured twice the pension for his wife than would otherwise have been the case. Even in the case of his own, he used death to demonstrate his superiority over the rest of the world.

Shipman, like many before him, almost got away because no one questioned his version of events. I'm a great believer in questions. Nothing is more important that satisfying yourself the

HAVE THEY BEEN USED AS MURDER WEAPONS?

information you are being offered is correct and I still believe in letting people ask any question in spite of some of those I was asked in the Poison Garden (which we'll look at in the next chapter).

9

Do You Mind Me Asking?

I love questions. I hate questions. For me, some of the most enjoyable tours I took round the Poison Garden were in the depths of winter, when there was very little to see, with, perhaps, as few as two or three people, because those were the times when people could ask whatever questions they wanted and you knew you were telling them things in which they had an interest. Contrast that with a summer tour, of over 40 people, with another tour following close behind and someone who won't stop asking questions even though you've made it as clear as possible that you don't welcome them in this circumstance.

Unfortunately, some of the people who insisted on asking questions, even after I'd said several times that I would answer them at the end of the tour, were the ones who didn't want to find information out; they wanted to show what they knew. 'Isn't henbane the poison that killed Hamlet's father?' they ask just so you understand that they are well read. In the circumstances of a busy tour, it's very tempting to just say 'Yes' and move on because the correct answer takes a lot of time.

Henbane or not henbane – that is the question

It is often said that henbane was the poison used to kill Hamlet's father in Act 1 scene v. Different versions of the text call the poison 'hebenon' or 'hebona'. The text talks of 'With juice of cursed hebenon' but goes on later to call it a 'leperous distilment'. It is this description that leads some scholars to propose that the poison must have been an extract from wood rather than simple plant juice. The Wikipedia entry for hebenon suggests that it was an extract of yew but Marshall Montgomery (1920) in *The Modern Language Review*, believes 'hebenon' derives from 'ebony' and was in fact Lignum vitae, which was thought to be a variety of ebony.

So, henbane was not, necessarily, the poison used to kill Hamlet's father. Shakespeare, generally, does not name the poisons he uses, which may be because he wanted to avoid the know-alls' questions about whether he'd got the right symptoms for the substance.

The other problem with questions from a large group was that, if the questioner was the only one interested in the point, the rest of the group could get bored and start to wander around. That was made worse if they couldn't hear the question. In that case, the least you should do is repeat the question clearly so everyone knows what was asked but it is better if you can find a way of repeating a question without appearing to.

But, if the circumstances were such that a question could be answered then they really brought a sparkle to a tour. Sometimes, mothers would be embarrassed about how inquisitive their offspring were but I would always assure them they should be pleased to have children who were willing to question what they were being told rather than just accept it. Because there is a lot of poison plant information around, which does need to be questioned and, I'm sorry to say, some of that came from my colleagues at the Alnwick Garden.

In Chapter 1, I showed that the story that the Romans brought Urtica dioica to Britain to beat themselves with to keep warm was highly unlikely to be true but it took a number of training sessions to get the guides to see beyond the story and realise it didn't stand up when questioned.

DO YOU MIND ME ASKING?

A box of poison

If you just assume something without question not only might you be misinformed but you may miss out on an interesting story. Take Buxus sempervirens, the box; many people assumed it was in the Poison Garden just as borders around the flame-shaped beds. They were, usually, surprised to learn that it is poisonous and, generally, amused to hear my reason for its inclusion in the garden.

Buxus sempervirens contains the alkaloid buxine, which makes the leaves poisonous to farm stock as well as humans. It causes nausea, vomiting and diarrhoea, and death may occur through respiratory failure, but its unpleasant odour and bitter taste tends to minimise its ingestion. There have, though, been instances of cattle breaking into gardens adjacent to farmland and grazing on a box hedge with unpleasant consequences. Most people who have any sort of problem with box do so as a result of handling fresh clippings, which can cause skin rashes.

This risk of skin problems from handling the clippings gave me a story, which managed to surprise me. I would explain to the visitors that, in addition to full-time employed gardeners, the Alnwick Garden was fortunate to have the services of a number of very skilled garden volunteers. I would then pretend that when the box hedges around the beds in many areas of the garden (not just the Poison Garden) were cut, an employed gardener would do the cutting, leaving a volunteer to collect up the clippings. Then one morning, before we got busy, I took a walk up to the Ornamental Garden to see what was in bloom and found an employed gardener cutting the box hedges and a volunteer clearing away the cuttings. At the risk of spoiling the story, I should say the volunteer was wearing a sturdy pair of gloves.

Though buxine is the main alkaloid present in Buxus, there are others and one of these, cycloprotobuxine, has been investigated as a chemotherapeutic agent in cancer therapy but, so far, has not gone beyond investigation. Trials by French workers seemed to show that an extract from box was helpful in reducing the amount of HIV virus in an infected person. This work was done before the present antiretroviral drugs were available and there is no indication that further trials are being undertaken.

In the French trial some subjects were given a low dose of the box extract, some a high dose and some a placebo. Those on the low dose responded well but those on the high dose exhibited some side effects of vomiting, diarrhoea, muscular spasms and paralysis.

So Buxus is poisonous and can justify its position in a poison garden but that isn't why it was used as a border round every bed. When talking about Hyoscyamus niger, in Chapter 8, I mentioned that it was a component of the flying ointment used by witches to induce the feeling of flying, especially for novices. This salve would be made up from a number of plant extracts including plants such as Atropa belladonna, Aconitum napellus, various types of Datura and mandrake. In other words, many of the plants in the Poison Garden. In his *Great Herbal* (1598), John Gerard mentions this salve. Gerard often tries to find a form of words that will not cause offence when describing plants whose use has a sexual connotation. For the witches flying brew, therefore, he talks about them applying it to 'their armpits and other hairy places'. It gives a whole different view of the reason witches sat on broomsticks in order to fly.

Now the thing about witches is that they like to know everything, which is one way they can exert their power over others. Most witches can tell you how many branches, twigs and leaves there are on any plant in the garden, except the box. You see, the leaves of the box are so small and the plant is so compact that whenever a witch tries to count them she is bound to lose her place and have to start again. You will often see a Buxus bush growing near the door of a house. It is there to prevent any witches from entering the property because, if one approaches, she will be found forever trying to count the leaves on the bush and forever having to go back to the start after losing her place.

DO YOU MIND ME ASKING?

So, the Buxus sempervirens around the beds of the Poison Garden were there to prevent witches coming into the garden to steal the plants they wanted to make up their flying ointment. Now, aren't you glad you asked instead of just assuming?

Some people are shy about asking questions. The first summer that the Poison Garden was open was 2005 and we had not recruited and trained enough guides to take everyone who was interested on a guided tour. It was decided to use what we called 'open guiding' to maximise the throughput. Members of staff were placed around the garden and visitors were encouraged to wander around and ask whatever questions they liked. The trouble was that, whoever was at the gates, would hear people leaving asking each other 'Why is such and such a plant in there?' while at the same time the guides in the garden were standing around not talking to anybody. Given that this was the year when the original planting was on display, we were aware that many people felt that calling it a Poison Garden was going a bit far when it was filled with marigolds, violets and fennel. Letting these people leave without having the chance to explain to them what the garden was about was likely to result in them not encouraging their friends to come and see us.

Knowing that many people will not come out and ask questions, I always tried to keep an eye out for anyone looking quizzical or muttering and, if the chance arose, I would try and answer their question even though it hadn't actually been asked.

I really need a cuppa

For some reason Camellia sinensis is a plant that attracted a lot of these unstated questions. People looked at the name Camellia and didn't realise that Camellia sinensis is the tea plant. Just explaining that was insufficient to satisfy their curiosity because they couldn't see why tea should feature in a poison garden.

IS THAT CAT DEAD?

Though there are still disputes about it, most people agree that tea was discovered in China, over 4,000 years ago when the Emperor Nun Shen, a scholar and herbalist, was kneeling beside a fire, boiling water. As the water boiled, a breeze blew the topmost leaves of a nearby tree into the pot. The aroma enticed Shen to taste the water, which he found delightful and claimed that this liquid was both delicious and invigorating. In India it is said that, Bodhidharma (Ta Mo), the eventual founder of Zen Buddhism, was part way through a seven year sleepless meditation when he chewed some leaves to keep himself awake and found they were refreshing. A more extreme version of the story says that Ta Mo was travelling in China in AD 520 spreading his teaching when he fell asleep. So ashamed was he at the teaching time lost by sleeping that he cut off his eyelids and threw them to the ground. The next day, two shrubs had grown where the eyelids fell and Ta Mo chewed the leaves and found them to be stimulating.

Tea contains caffeine and, though coffee is more often thought of as the source of this addictive substance, expert opinion suggests that five cups of tea a day is enough to produce addiction. The effects of caffeine addiction are, often, underestimated because it challenges the general view of what being an 'addict' means. But the physical effects of caffeine withdrawal are well documented and can be similar to withdrawal from tobacco or heroin. A number of visitors' stories confirmed that withdrawal from caffeine can produce physical symptoms, which, though not usually as severe as withdrawal from these other substances, are unpleasant at the time.

A visitor described what happened when he returned home after two weeks working away during which he drank 14 cups of tea a day. He said he had 'the worst hangover' he'd ever known with headaches, stomach upsets, sleepless nights, heart palpitations and general debility. And a number of visitors said they had experienced the typical 'weekend headaches' resulting from consuming much more tea or coffee during the working week than when at home.

DO YOU MIND ME ASKING?

The effect of caffeine was illustrated by a visitor, a clinical pharmacologist, who said she had been involved in a drug trial in which one of the patients produced a completely different reaction to all the others. It turned out he had been consuming huge amounts of coffee and the caffeine had affected the action of the drug on trial.

Anyone who has ever had to have surgery under general anaesthetic knows that the first couple of days after surgery can be a fairly miserable time in terms of general health rather than pain associated with the procedure. In a paper given at the 2002 International Symposium on the History of Anaesthesia, M. van Wihje notes the effects of caffeine withdrawal and suggests that this might be the cause of some of this post-operative discomfort. He thinks giving caffeine post-operation might alleviate these effects. I've found no indication of anyone conducting any controlled trials on this possibility but my advice to anyone awaiting surgery is to cut down on the caffeine for a couple of weeks beforehand.

The introduction of substances like tea, coffee, chocolate and tobacco into Europe was, often, influenced by economic, political and social factors. In some parts, their adoption was opposed by the authorities because it was felt that men gathering together to drink or smoke could be a cover for political dissidents to meet. In others, the interests of local alcohol producers were defended by official opposition to the new beverages. (For some time smoking was referred to as having a drink of tobacco.)

Consumption was sometimes encouraged by claims of medical efficacy though such claims may have been based on the desire of the importers to boost sales. Cornelis Bontekoe was a member of the 17th-century Dutch empirical medical school, which sought to use concepts of healthy and unhealthy to overthrow the belief in the 'humours'. He was, later, as private physician to Frederick William of Brandenburg, one of the key people in bringing coffee to German society. Prior to that he had, whilst rumoured to be in the pay of the Dutch East Indies Company, said that drinking eight to ten cups of tea a day was the minimum required to obtain its health benefits, but he saw no reason not to drink up to 100 cups a day!

Scientific promotion of the benefits of drinking tea still occurs in our times. A recent research report said that drinking three or more cups of tea a day is at least as good for you as drinking a lot of water and may even be better. The researchers denied that their results were in any way affected by the fact that the Tea Council funded the work amid concerns that young people are more inclined to drink bottled water than tea.

Not everyone has always shared the view that caffeine is good for you. In the 18th century in Germany, a ruler insisted that his people should drink beer at breakfast time instead of coffee, so alcohol was preferred to caffeine.

Children should be...

Sometimes, my wish to encourage children to develop the habit of questioning everything had unforeseen consequences. At the end of a tour, I was talking to a woman as we left the garden and her son, about four or five, obviously had a question he wanted to ask. His hesitant, puzzled look showed that he really wanted to know whatever it was that was on his mind but he wasn't sure he could ask. I told him he looked as though he had a question on his mind and that it was perfectly all right and he could ask me anything. He looked up at me and said, 'Why have you got such a big red nose?'

On another occasion, a late tour started with a single group comprising two mothers and about four children, one of whom, a little boy, started asking lots of questions. After a few minutes, we were joined by six adults so I stopped answering his questions. He started putting his hand up so, from time to time, I would ask him what his question was. We'd completed the tour and were down by the cannabis when his hand went up. I said, 'One last question before we go' and he started: 'One night, when I was in bed and mummy was on the toilet...' That was as far as he got. His mother scooped him up and pressed his face to her chest so all we could hear was him mumbling, and ran off up the tunnel, soon followed by the other mother and the rest of the children. One of the six adults said, 'Now we'll never know.'

I think there was only one question I dreaded and then only if it was asked at the wrong time. I had developed a routine for dealing with one of the plants, which built up to a climax and a payoff that usually went down very well. But it could be completely spoiled if someone, before I got into my stride, asked, 'Does mandrake really scream?'

10

Does Mandrake Really Scream?

There's something special about Mandragora officinarum, mandrake. For a start it used to be called Atropa mandragora, emphasising both its relation to the deadly nightshade and its connection with *Atropos* the Greek Fate who held the shears that could be used to cut the thread of life. Because, mandrake is a potentially fatal poison. But that does not explain man's extreme fascination with this plant; a fascination that means entire books have been written about this one plant and its associated folklore. Incidentally, I deliberately used the words 'man's extreme fascination' because, as we will see, this plant has been of more interest to men than women for most, though not all, of its history.

Origins

The first known mention of mandrake is disputed. Some sources, even today, maintain that it was known and used in ancient Egypt. This belief that the Egyptians knew and used mandrake root is based on Dr Joachim's German translation of the Egyptian hieratic word *d'd'* or *didi*. Joachim took this to be from the same root as the Hebrew word *dudaim*, which is the mandrake. However, between Joachim's late 19th-century German translation and Dr Bryan's English version, written in 1930, several eminent Egyptologists had concluded that *d'd'* was in fact a mineral, haematite. The main evidence of use of mandrake in Egypt is, therefore, suspect. What remains are designs on various artefacts from Egyptian times, which do appear to be stylistic depictions of the mandrake. It is possible that the Egyptians knew of mandrake as an imported extract and the depictions of it were based on descriptions of the plant rather than first-hand observation.

For most people, the agreed first mention of mandrake is in the Old Testament. Jacob was married to Rachel but had slept with Leah, her elder sister, on the wedding night. Leah produced children for Jacob but Rachel was barren, a cause of conflict with Jacob and great distress to her.

> And Reuben went in the days of wheat harvest, and found mandrakes in the field, and brought them unto his mother Leah. Then Rachel said to Leah, Give me, I pray thee, of thy son's mandrakes.
>
> (Genesis 30: 14)

At this time, it seems, mandrake was favoured as a cure for female sterility rather than a male aphrodisiac though it must be considered a possibility that Rachel's problem was that she could not arouse Jacob sufficiently to become impregnated and, perhaps, thought a little help from mandrake would solve Jacob's lack of passion towards her.

There is a mass of speculation about whether the plant found by Reuben was mandrake and whether it is the same plant described in The Song of Solomon:

> The mandrakes give a smell, and at our gates are all manner of pleasant fruits, new and old, which I have laid up for thee, O my beloved.
>
> (The Song of Solomon 7: 13)

Arguments centre over whether the 'smell' was pleasant, since the smell of mandrake is not, at least not to European noses. Some people say that mandrake would not be ripe at the time of wheat harvesting, which others counter by saying that Genesis does not say the plant was ripe. And so it goes on.

We can learn something of interest by turning away from the biblical text and considering the original Hebrew folktale on which the story is based. In this, Reuben did not recognise the mandrake but just happened to use it to secure his donkey to while he was working in the fields. When some distance away, he heard a terrible scream and found his donkey dead and the mandrake pulled out of the ground. From this tale comes the long history of the belief that mandrake screams when pulled from the ground and that the nearest creature, hearing the scream, will die. Rationally, it may be that the donkey was spooked by a snake or other creature and, in attempting to flee, choked itself on the rope securing it to the plant and the scream Reuben heard was that of the donkey.

Whatever the truth, and we shall never know, the story of the mandrake screaming passed into folklore throughout Europe and the East and combined with mandrake's other special attribute to build a wealth of tales and myths.

As aphrodisiac

Mandrake has a bifurcated root, that is to say the root, typically, splits into two and looks like the legs of a human being. Now, many root vegetables such as carrots and parsnips will, often, grow in this way but it is the mandrake that became famous for this property. Under the Doctrine of Signatures (the idea that the look of a plant told you how to use it) mandrake would obviously make a man a man.

In fact, mandrake was used in ancient Greece as an anaesthetic for operations. The patient was given the root to chew on. Larger doses could be expected to result in respiratory failure though it is said to produce vomiting and diarrhoea if larger amounts of the root itself are ingested. There seems to be nothing to justify its reputation as an aphrodisiac. It is always tempting to see our ancestors as ignorant fools and use this to explain how a plant with no chemical effects on the reproductive system could retain its reputation for over two thousand years but this is to underestimate the intelligence of our forebears.

Within the last ten or so years, a condition described as stress impotence has been identified and is accepted as a genuine cause of arousal problems by at least some of the authorities on this subject. The theory is that the pressure to perform well in bed can lead some men to either fail to achieve erection or to ejaculate prematurely. It seems possible that use of a small amount of anaesthetic (mandrake) could produce sufficient relaxation to remove this stress and, coupled with a strong placebo effect based on its well-known efficacy, result in enhanced performance for the user.

We can be pretty certain that the amount used would have been small since mandrake was always expensive. This would have been due to its comparative rarity in northern Europe, where it struggles to grow to any size, plus the great difficulty in harvesting it.

Harvesting and handling

The need to avoid the risk of dying when pulling mandrake up led to many different ways of harvesting it (without having to endure the scream). What follows is just a summary of the more common beliefs about mandrake and the gathering of the plant.

The simplest and most widespread method for obtaining mandrake is to tie a dog's tail to the plant, retreat a safe distance, stop up your ears and call the dog. Variations of this method include starving the dog for several days and throwing a piece of meat to entice it to pull on its tether, using only a black dog and loudly reciting various incantations to block out the sound of the screaming plant.

DOES MANDRAKE REALLY SCREAM?

But not all cultures require the use of a dog. In parts of Eastern Europe, it was essential that only an expert attempt to uproot a mandrake as any damage caused to the root would be replicated on the person who caused it. In Armenia, the mandrake was to be visited on three successive days in the company of a young and handsome virgin prior to being harvested. The idea of a, probably, lusty young man, going into the woods to visit a plant with the alleged properties of mandrake in the company of a young, handsome virgin makes one wonder whether, by the third day, the young woman still fulfilled the required criteria.

Josephus Flavius, writing in the first century, says that mandrake withdraws when approached adding to the difficulty of pulling it up. The way to keep it on the surface of the ground so it can be extracted is to pour over it urine or menstrual blood. He says touching it is certain death unless you can be sure of taking the entire root. There is no mention of screaming but he does suggest using a dog to pull it out of the ground. It is recommended for driving out demons. Just bringing it close to a possessed person is enough to drive the demon away.

The *Apuleius Platonicus* is an Anglo-Saxon herbal believed to have originated in about AD 400. With constant copying, it is likely that the original was amended over the centuries. The so-called 'Cotton' manuscript, dating from the first half of the eleventh century, echoes Josephus, but says that mandrake will fly away when approached. It says you should surround it with iron to stop it flying away but also states that mandrake shines at night like a lamp, making it easier to find in the dark.

The belief that mandrake grew under a gallows was widespread but had numerous refinements. In Wales and parts of Germany, it grew from the tears of an innocent man, hanged. Elsewhere in Germany, the requirement was that the hanged man be the son of a family of thieves or someone whose mother stole when pregnant. Thus, the hanged man had not chosen to be a thief but could not avoid the family business and so retained some degree of innocence. Sometimes, the hanged man had to have never had sexual intercourse. Generally, bodily fluids other than tears were what caused the mandrake to grow.

Not every culture required mandrake to be consumed to achieve beneficial effects. In many countries, but especially in Germany, mandrake was kept as a talisman, especially if the root bore a particularly striking resemblance to a person. Often mandrake was uprooted, cut to make the appearance more human and replanted so that the cuts would heal and the shape would appear entirely natural. Later in the chapter, we'll look at how other plants were made to look like mandrake and what problems that could cause.

In 15th-century Germany a root with a strong resemblance to a man was highly prized and such roots were often kept in small wooden boxes, being passed down the generations as family heirlooms. There are still examples of such roots in many museums in Europe and the Middle East. It was, however, essential to take the manikin out of its box on a Friday and give it a bath. Failure to do this would cause the root to shriek until it was bathed. This wasn't a completely worthless chore since the bath water could be sold to be drunk by a pregnant woman, or rubbed on her stomach to provide pain free childbirth.

Many people carried mandrake root about their person for luck. Alexander the Great is said to have carried a mandrake root with him at all times. It has been suggested that Alexander the Great's death was the result of a mistaken belief in the power of plants. Until relatively recently, a common treatment for diarrhoea was to give a strong laxative, the theory being that whatever was causing the upset could be 'purged' from the bowels. It is thought that Alexander insisted that his surgeons continue to give him a potion made from Veratrum album, white hellebore, even though his condition was deteriorating.

DOES MANDRAKE REALLY SCREAM?

The sceptical herbalists

In general, the herbalists were not subscribers to the folklore of mandrake. Pliny the Elder ignores much of the folklore, especially any reference to screaming, but notes that:

> The diggers avoid facing the wind, first trace round the plant three circles with a sword, and then do their digging while facing the west.

He goes on to say that the juice of leaves that have been touched by dew is deadly. Even keeping the leaves in brine does not remove the harm. The mere smell can make the head feel heavy and smelling it too much can leave you struck dumb. Used as a sleeping draught, the strength of the patient should be considered when judging the dose. It can be ingested or smelt to produce anaesthesia.

Pounded with pearl barley, mandrake treats gout and eases joint pains and is used against erysipelas, a skin infection resulting in a burning pain mostly in the face. It has a number of gynaecological uses from delivery of a stillborn baby to reducing excess menstrual bleeding. Pliny also says it can be used to cure superficial abscesses but, as the alternative is to use a poultice of Verbascum applied by a naked maiden, it was probably not widely employed in this role.

Dr William Turner (1551) dismisses many of the stories associated with mandrake and its alleged talismanic benefits:

> The roots which are counterfeited and made like little puppets and mammets, which come to be sold in boxes, with hair and such form as a man hath, are nothing but foolish feigned trifles, and not natural; for they are so trimmed of crafty thieves to mock the poor people withal, and to rob them both of their wit and their money. I have in my time, at divers times taken up the roots of Mandrag out of the ground, but I never saw any such thing upon or in them

as are in and upon the pedlars' roots that are commonly to be sold in boxes... But it groweth not under gallosses [gallows], as a certain doting doctor of Cologne in his physic lecture did teach his auditors; neither doth it rise of the seed that falleth from him that is hanged; neither is it called Mandragoras because it came of man's seed, as the foresaid doctor dreamed.

He does, however, go on to say that mandrake poisoning can be cured by pouring a mixture of rose oil and vinegar over the head of the victim. Turner also talks of mandrake's use as an anaesthetic in his own time.

John Gerard (1598), after noting that the division of the mandrake root is no different from that often seen with carrot or parsnip and such like, goes off on one of his rants about the tales associated with the plant:

> There have been many ridiculous tales brought up of this plant, whether of old wives, or some runnagate surgeons or physicke-mongers I know not, (a title bad enough for them) but sure some one or moe that sought to make themselves famous and skillfull above others, were the first brochers of that errour I speake of. They adde further, that it is never or very seldome to be found growing naturally but under a gallowes, where the matter that hath fallen from the dead body hath given it the shape of a man ... with many other such doltish dreams... Besides many fables of loving matters too full of scurrilitie to set forth in print, which I forbeare to speak of. All of which dreames and old wives tales you shall henceforth cast out of your books and memory; knowing this, that they are all and everie part of them false and most untrue: for I my selfe and my servants also have digged up, planted, and replanted very many, and yet never could either perceive shape of man or woman.

DOES MANDRAKE REALLY SCREAM?

At the end of this rant, Gerard goes back to his more usual style with a detailed description of the female plant but he is one of those to question the Genesis story of the mandrake, maintaining that the plant Reuben brought home was not mandrake because the Hebrew should be translated as 'amiable and sweet smelling flowers', which the mandrake does not have.

Leonhart Fuchs (1501–1566) published *De historia stirpium* in 1542. The book was an attempt to bring together text and illustrations of medicinal plants, mostly taken from ancient Greek physician and botanist, Dioscorides. After this he began work on a much more comprehensive encyclopaedia of plants but died before it could be published. Much of the preparatory work is available in the form of text and illustrations.

On mandrake he says:

> Mountebanks and fakers hanging around the marketplace are peddling roots shaped in human form they claim are Mandragora although it is quite evident that they are fashioned and made by hand from Canna roots carved in human likeness.
>
> (Fuchs, 1542)

Canna roots, a commercial source of starch, were just one of the plants used to produce fake mandrake. The one that most often seems to have served this purpose is Bryonia dioica, the white bryony, sometimes called 'English mandrake' because its substitution was so common. Bryonia has a very vigorous root, which can grow to a prodigious size. John Gerard (1598) says that the Queen's surgeon, William Goderous, showed him a root weighing half a hundredweight and the size of a one-year-old child. This rapid and copious root growth meant that the 'mountebanks and fakers' could plant a bryony in a mould shaped like mandrake root and it would quickly fill the mould and take the required shape. Obviously, as the leaf growth is completely unlike mandrake, these roots would be stripped of their foliage before being taken to market. Alternatively, a plant could be dug up, carved to shape and replanted for long enough for the cut edges to grow over. Either way, a reasonably quick turnaround could be achieved because of the rapid growth of the root.

Given that the action of the true mandrake root was, probably, largely due to a placebo effect based on the reputation of the plant, there seems to be little harm in passing off bryony root as mandrake. There is an argument that says that if a sugar pill is as effective as a £5 per pill pharmaceutical, which is often the result found in double blind controlled trials, why not prescribe the sugar pill but say it is the expensive drug; the effect on patients will be the same. With the bryony, however, there is a downside to that argument; it is a dangerously strong laxative. Even in the tiny amounts believed to have been consumed, bryony would be likely to have produced strong purging. One wonders whether those who had been duped returned to the market to complain only to be told that the seller had only promised that anyone eating his product would be up all night.

Fiction and theology

With such a rich folklore, it is inevitable that mandrake has been widely employed in works of fiction. It is noticeable, however, that most of these refer to its ability to produce sleep, often a deathlike sleep. Although not referred to by name, Chaucer, in *The Knighte's Tale*, talks about a wine made of narcotics, which made the jailer sleep. Shakespeare has a number of references to mandrake either directly or by implication from the unnamed potion employed in the plot. In *Henry IV, Part 2*, Falstaff twice refers to mandrake, comparing Justice Swallow, naked, to the root. In *Antony and Cleopatra*, Cleopatra asks for Mandragora to help sleep through the time Antony is away. In *Othello*, Iago says that the Moor's jealousy is preventing him from sleep and neither mandrake nor opium could help. Some sources think Juliet took mandrake to induce her deathlike sleep and this power is also alluded to in *Cymbeline*. The shriek of pulling it up is mentioned in *Romeo and Juliet* and *King Henry VI* and its power to cause insanity comes into *Macbeth*.

But it is not just Shakespeare. Ben Jonson in *Masque of Queens* has a witch talk of gathering the root. Marlowe in *The Jew of Malta* effects Barabas's escape from prison by having him drink mandrake so that he is thought to be dead. Webster in *The Duchess of Malfi* makes several references to mandrake. Though not mentioned by name, some have argued that the 'love potion' in Wagner's *Tristan and Isolde* was based on mandrake. There are many other references in less well-known works and there was even a silent film, *Alraune*, made in 1928, where mandrake growing at the feet of a hanged man played a major role in the plot.

Mandrake has also been used in more modern works of fiction. One episode of the BBC TV series *New Tricks* involved dogs being poisoned by being given fillet steak impregnated with mandrake root. This led to a red herring suspect in the form of an Egyptologist because of 'the importance of mandrake root in ancient Egypt'. In *The X-Files* episode *Terms of Endearment* (series 6, episode 6), a woman is found dead from mandrake poisoning apparently after trying to induce an abortion.

But, of course, the best known fictional use of mandrake, in recent times, comes in the Harry Potter series by J.K. Rowling. In *Harry Potter and the Chamber of Secrets*, the young wizards are taught that mandrake will reverse petrification, a condition where a person is apparently turned to stone. To avoid the upset of having dead dogs around the place, Rowling says that young plants can be safely pulled up when wearing stout earmuffs and that the scream of the young plant will only cause unconsciousness, not death.

It is interesting that Rowling uses mandrake to reverse the turning to stone because this alleged ability to produce a deathlike sleep leads to another area of speculation and controversy, which persists to this day. Those who try to rationalise the resurrection of Jesus Christ claim that the sponge which he was offered when he cried out that he was thirsty was soaked in an extract of mandrake so that he would appear to have died and could be hidden away to reappear as if risen from the dead. This argument is strengthened by the apparent departure from normal practice during his crucifixion in that his legs were not broken, which was the normal procedure to speed the death of those being executed in this way.

So the power of the mandrake tells us that counterfeiting goods is not something that started with fake perfumes or battery-powered Rolex watches. It helps to add to the appeal of children's books and it still provokes debate about the central tenet of Christianity.

11

Why Don't We Buy the Poppy Crop in Afghanistan?

Comments and questions were always a very welcome part of my role as Poison Garden Warden at the Alnwick Garden and, generally, the tone of them was light-hearted reflecting the fact that visitors were on a day out for enjoyment and hadn't come to the garden to get a lecture. But one comment struck me deep and I can still remember going cold when I heard it.

I would always bring my tours to a climax with the Cannabis sativa. For most people, this was their first sight of a growing cannabis plant and, for the others, you could easily spot the ones who knew what they were looking at and looking for in the plant. My position on illegal drugs is fairly straightforward: what matters is to understand them and reach your own decision on what your approach to them should be without letting those around you bully you into doing something you don't want to do or lie to you about their effects.

Though our ancestors told many lies about plants, which were used to keep children away from harmful plants, I don't believe that lying about the effect of substances is useful today because young people have so many more sources of information than were available hundreds of years ago. Take cannabis, for example. There are those who claim that cannabis use leads on to the use of 'hard' drugs and there are studies that show that a high proportion of addicted drug users began by using cannabis, and this is used to argue that cannabis is a 'gateway' drug. The problem with that argument is that it may be an example of the principle of *post hoc ergo propter hoc*, a later event happened because of an earlier event, whereas there is no evidence for a causal link.

My own view is that telling lies about cannabis may be the mechanism which makes it a 'gateway' drug. If young people are told that cannabis is extremely harmful and that use will lead to idleness, aggressiveness and mental health problems they are highly likely to find their own experience and other sources of information available to them to be markedly different. If they are now told that heroin is an addictive substance, which leads to many deaths each year due to overdose, they may look at the situation regarding cannabis where they were lied to about its effects and assume that what is being said about heroin, or any of the other 'hard' drugs, is also lies.

So I tried, during my tours, to set out, simply, and without taking all of the 20-minute tour to do so, the potential for harm these substances (and the legal ones like tobacco and alcohol) have if misused without overstating the points. During one tour, as I was talking about the harm done by these substances, I noticed a woman become upset and rest her head on the shoulder of the friend who was with her. I wasn't sure if it was something I'd said or if she was just feeling unwell, so at the end of the tour I went to ask if everything was OK. She said it was and thanked me with the words 'If my son had heard you talk, three years ago, I don't think he would be dead now'.

In Chapter 4 we looked at the number of deaths due to heroin overdose and, in Chapter 8, I gave examples of its use as a murder weapon. In this chapter, we'll look at the growing of Papaver somniferum and the resulting production of morphine and heroin and discuss some of the issues surrounding the measures to limit its availability and the harm it causes.

Papaver somniferum is a plant that grows readily in many parts of the world. Historically, India was a prime source of the opium

WHY DON'T WE BUY THE POPPY CROP IN AFGHANISTAN?

exported by the British to China and, as we've already seen, the plant was known to the ancient Egyptians. From the end of the Second World War until the end of the 20th century, 'The Golden Triangle' was the main source of illicit opium poppies. The 'Golden Triangle' is a remote mountainous area of approximately 100,000 square kilometres roughly centred on the boundaries of four countries: Burma (now Myanmar), Thailand, Laos and Vietnam, where two rivers, the Mekhong and Ruak, meet. According to the United Nations Office on Drugs and Crime (UNODC), in 1991 there were over 200,000 hectares of opium poppies growing in the Golden Triangle but by 2006 this had fallen to 24,000 hectares and it is expected that the area could become opium free within a few years. Most of this fall has come since the start of the 21st century though a 1996 agreement between Khun Sa, one of Shan State's most powerful drug warlords, and the ruling junta in Myanmar is thought to have started the decline.

All governments in the region also acted to clamp down on production and provide alternative sources of income, but it is likely that the massive increase in output from Afghanistan may, simply, have made it harder for opium growers in the Triangle to find a market. It must also be said that, to a large extent, Myanmar has gone from growing opium poppies to manufacturing methamphetamine so the area has not completely severed its connection with the illegal drugs world. The latest UNODC 'World Drugs Report' also says that opium cultivation in Myanmar increased by 29 per cent in 2007 so continued efforts are required if the area is not to revert to the trade. Meanwhile, there are at least two museums in Thailand dealing with the history of the Golden Triangle so it is to be hoped that 'history' is the right description for the area's involvement in the heroin trade.

In the same period, which saw the fall in output from South-East Asia, Afghanistan production was going in the reverse direction. The area under cultivation went from 41,000 hectares in 1990 to 193,000 hectares in 2007, though estimates for 2008 suggest a drop to 157,000 hectares. The trend has been upward throughout this period with the exception of 2001 when less than 10,000 hectares were planted, after a Taliban directive, in July 2000, banning the growth of the poppy.

There has been much debate about this ban. It is of note that the ban was only on the growing of the opium poppy and not on the trade in opium or heroin. In fact, the ban resulted in a tenfold increase in the price of heroin and, thus, produced a short-term upsurge in the income earned from heroin much of which, most people agree, finds its way to the Taliban. As well as a way of boosting its income in the short-term, in a way not unlike OPEC's cutback in oil production in the 1970s boosted the income from the oil that was produced, a ban on the growing of opium was, perhaps, a way for the Taliban to demonstrate its authority in the country.

It may be that, having caused a massive increase in the income from opium, the Taliban would have permitted some opium poppy growth under strict control so as to maintain prices but that can never be known due to the invasion of Afghanistan and their overthrow from central government. Some critics of the invasion have suggested that there were under the counter deals allowing the growth of the poppy in return for support for the new regime but the fact remains that the opium ban issued by President Karzai on 17 January 2002 was the first to completely outlaw the heroin trade.

That there have been massive and rapid increases in poppy production in areas where the Taliban have most influence certainly suggests that they were not, as some people have suggested, fundamentally opposed to opium on religious grounds.

In 2007, Afghanistan accounted for 82 per cent of the total area under poppy cultivation in the world but, because of climate, variety and growing conditions, the yield from Afghan poppies is higher than for other areas and, thus, 92 per cent of opium, by weight, came from Afghanistan's growers. That situation, naturally, raises concern for governments anxious to win the 'War on Drugs' but almost everyone, including the commanders themselves, recognises that military action will never eliminate the Taliban and, without order and good government, you cannot expect to eliminate the opium poppy.

This recognition has led some voices, including UK national newspapers, to call for the purchase of the Afghan poppy crop so that it can be converted to morphine, codeine and diamorphine for legitimate medical use, thus overcoming the long-running shortage of these essential painkillers. Sadly, that option is not as easy, nor as effective as it appears.

Saying that Afghanistan grows 92 per cent of the world's opium is only part of the story. A very small part of Afghanistan grows 92 per cent of the world's opium. Eighteen of the country's 34 provinces grow no poppies at all and, of the 16 which do grow the poppy, nine account for only 2 per cent of the total output. By contrast, Helmand provides over two thirds of the total opium produced. Some areas are poppy free because they cannot grow the poppy and throughout the country quite a number of farmers refuse to grow the poppy on religious grounds. But that leaves substantial areas that could grow the poppy but do not, at the present time, because they cannot compete with the areas that do.

So, the first problem with buying the existing crop is that other areas would move into poppy cultivation to fill the gap in the market. Then there is the political problem arising from appearing to reward those who break the law at the expense of those who observe it. Paying only those farmers who currently grow poppy would run the risk of alienating those who could but don't and, possibly, lead to greater support for anti-government forces. It would, therefore, not be enough to pay the estimated $1 billion that farmers make from opium. Much higher amounts would be required to compensate all potential growers to the same level.

As to the argument that the opium bought could be used to produce legal pharmaceuticals and, therefore, it is not money wasted, that completely misses the relative market sizes. Opium is used to produce morphine, which can be used as it is or is further processed to produce codeine and diamorphine. It is true that there is a shortage of, particularly, diamorphine because not enough poppies are being grown in places like Tasmania, Spain and the UK to meet the demand from hospitals and GPs, and it is true that this shortage produces genuine suffering. A number of visitors to the Poison Garden talked about their own or their relatives' experiences of having to wait before beginning a course of cancer treatment because the hospital could not guarantee the required supply of diamorphine to complete the course and this was confirmed by doctors and other hospital staff. One hospital pharmacist demonstrated a bizarre effect of this shortage. Because doctors cannot guarantee continuity of supply of diamorphine they are adopting other, less effective, treatment plans to avoid the need for it. This resulted in the situation that, when supplies of diamorphine were received, they sat on the pharmacy shelves until they went out of date and had to be scrapped.

So, there clearly is the need for an increase in the supply of diamorphine, but by how much is the issue. Total illicit opium production in 2007 was just under 9,000 tonnes whereas legitimate growers produce poppies that yielded the equivalent of around 500 tonnes of opium. Purchasing the whole of the Afghan crop would lead to a massive oversupply in the pharmaceutical industry and put farmers who have supplied this sector out of business. They would, thus, also need to be compensated.

But the greatest problem is that the opium industry in Afghanistan does not end with the growers. As already mentioned, it is estimated that growing opium gives farmers an annual income of $1 billion but the total value of the opium/heroin trade to Afghanistan is put at $2.5 billion, the additional $1.5 billion coming from processing and transport. The people involved in this trade can hardly be expected to sit back while their livelihood is taken away and would, beyond all doubt, act rapidly to find replacement growers to furnish their raw materials. The cost of buying out the opium crop just went from the $1 billion dollars often mentioned to $2.5 billion.

WHY DON'T WE BUY THE POPPY CROP IN AFGHANISTAN?

Even if it were decided that this sum was justified, it could not be spent. The majority of the $1.5 billion 'added value' between the grower and the heroin leaving the country goes into the pockets of anti-government forces generally grouped under the title 'Taliban'. They use this money to finance their insurgency including the purchase of arms.

In other words, if western governments contributed to the $2.5 billion dollars a year intended to stop the growth of the opium poppy they would, in effect, be paying the Taliban to kill NATO troops.

But the picture is not entirely bleak. Opium production in Afghanistan jumped over 50 per cent from 2003 to 2004 and has continued to grow since then. In 2006, output jumped nearly 2,000 tonnes but the United Nations surveys indicate that the demand for heroin has been broadly stable for some time. At the time of writing, in the autumn of 2008, approximately 6,000 tonnes of opium, which amounts to about 600 tonnes of heroin, is unaccounted for. What no one can say is whether this is being deliberately withheld to stop a complete collapse in the price or whether the Taliban is stockpiling it, believing that a time may come when efforts to prevent cultivation are effective or, and here is the glimmer of hope, there is a more or less finite demand for heroin in the world.

If this last is the case, then the proper course for governments to take is to focus on the demand side of the market: increase the availability of treatment to help problem drug users bring their lives, and their habit, under control; and concentrate on meaningful education of young people to reduce the number of new heroin users. Eradicating production is an impossibility since there are so many places where poppy is not being grown but could be and will be as long as demand exists but, if current demand is finite, then the problem of controlling demand is also finite and the right actions may be able to minimise the demand and support those for whom addiction to heroin is their lot in life.

IS THAT CAT DEAD?

It is time to ask whether criminalising those who use substances with the power to produce dependence, addiction and great harm is the best way to deal with the willingness of parts of the human race to use these substances, a willingness, which as we will see in Chapter 15, goes back thousands of years.

12

Should Local Councils Grow the Castor Oil Plant, Ricinus communis?

A common reaction from visitors as they left the Poison Garden at the end of their tour was one of amazement mixed with concern that 'I've got most of those plants in my garden!' And, in today's increasingly risk averse world, some of them would immediately start talking about removing some of or all of the offending plants 'because the (grand)children play in the garden'. My initial response to this would be to point out the same plants were in their own grandparents' garden and they had come to no harm and also to mention the number of very poisonous plants that many local authorities plant in public areas every year.

Ricinus communis, the castor oil plant, is what gardeners call a 'dot' plant. In the United Kingdom it is an annual, which grows to several feet high very quickly and, therefore, creates the plant equivalent of a punctuation mark when planted with other low growing annuals like marigolds. Before we look at why this is not a problem, it's worth taking a brief diversion into the subject of common names.

What's in a name?

We would often be asked why the name plates in the garden only had the botanical names of the plants and not the common names. Incidentally, it is not strictly correct to call them the 'Latin' names because that makes many people assume the names were given to the plants in Roman times whereas the current system for naming plants was developed by Carolus Linnaeus (1707–1778) who began publishing his works on the classification of plants with a slim pamphlet in 1735. Over his life this grew into several volumes and became the agreed standard by which plants were grouped into the family, genus, species, variety scheme still used today.

Linnaeus based his naming system on the names that many plants had had since Roman times but, where necessary, he 'created' Latin words, often basing them on the Greek name for a plant, to complete the classification. Many of today's plant names would be unrecognised by someone like Pliny the Elder or Dioscorides. So, it is more correct to talk of a plant's 'botanical' name rather than its 'Latin' name.

The great advantage of the Linnaean system is that a plant can be uniquely named and that was one of the reasons for using only the botanical names on the small metal signs next to the plants. Common names can produce confusion because they get applied to more than one plant, and not just different species or varieties within the same genus. For many people 'Granny's Bonnet' is some form of Aquilegia, though the term gets applied to many different species, but for others the name refers to Isotropis cuneifolia. 'Jack-in-the-pulpit' is the name sometimes given to Arum maculatum but more usually applied to Arisaema triphyllum. In both cases the name is satirical. The plants produce a long, erect spadix wrapped around by a cloak-like spathe. The inference was that church ministers had become remote from the common people and had inflated opinions of their own worth. The spathe was, therefore, the very grand pulpits erected to ensure the minister would be warm in a cold, draughty church, and the only part of the minister visible to the ordinary people was his over-large erection. The name 'Jack-in-the-pulpit' was, effectively, the same as 'priest's pintle', where pintle means penis. This was

SHOULD LOCAL COUNCILS GROW ... RICINUS COMMUNIS?

changed to 'cuckoo pintle' and, then, 'cuckoopint'. These days, many people are not aware of the origin of the name and say 'pint' as in a 'pint of beer' rather than saying 'pint' to rhyme with 'mint'.

As well as there being many instances of overlap between common names, many plants have several of them, often the result of regional or country variations. With visitors coming from all over the world, trying to show common names would have made the signs dominant in the garden. Dilston Physic Garden is a small garden in the north-east of England where staff are not always on hand to answer questions about plants. For that reason, the plants each have a sign giving the botanical name, the most common common names and brief details of the medicinal or poisonous uses of the plant. As a result, the garden itself is hard to enjoy because it looks more like a collection of advertising hoardings.

With Ricinus communis, the confusion over common names results from the reluctance of some garden centres and nurseries to deter potential buyers. The leaves of Ricinus communis bear a remarkable resemblance to those of Fatsia japonica. Not all varieties of Ricinus communis produce the red leaves, which some people think is the characteristic way to distinguish the plant and though, when Ricinus communis does produce green leaves these are less shiny than those of Fatsia japonica, there can be confusion between the two. Fatsia japonica is usually called 'false castor oil plant' to separate it from the 'castor oil plant' but some sellers are reluctant to label a plant 'false' because of its negative connotations. This can lead, as we know from hearing the story from more than one visitor, to someone ending up with the wrong plant. Having seen a friend's Fatsia japonica and been told it is the castor oil plant, people have gone to a garden centre and purchased a castor oil plant, Ricinus communis.

When these people realised the confusion during their visit to the Poison Garden, it would often need a little time to convince them that this was not such a cause for concern.

Laxative and lubricant

Ricinus communis is capable of producing one of the most toxic substances known to the human race, ricin, which is a simple protein. It also contains Ricinus Communis Agglutin (RCA), which causes increased coagulation and has led to confusion over the effects of ricin. Visitors have said they thought ricin stopped the flow of blood. In fact, ricin causes vomiting (which in many cases expels the poison and prevents death), stomach pain and dehydration and then destroys the main internal organs. With no antidote, treatment relies on symptomatic support and the use of heart/lung machines, kidney dialysis, etc. If the patient survives three to five days, symptoms usually diminish.

Oil made from the beans of Ricinus communis has been widely used as a laxative and tonic for children since about 2000 BC in Egypt. The ancient Egyptians were familiar with the 'castor oil tree' and made extensive use of castor oil and other parts of the plant in the remedies collected in the Ebers Papyrus, one of the two oldest known preserved medical documents. It appears in over 100 prescriptions. As might be expected it was drunk as a cure for constipation but it was also rubbed on the head to cure headaches or to promote growth of hair for a woman. In these instances it is used on its own but it also gets mixed with all manner of other items. With yeast and water, it would cure roundworm. With red lead it was capable of curing herpes on the face. The leaves, mixed with honey and 'clay from a statue' and applied as a poultice would treat a 'flow of matter' from both eyes. The leaves have been used in the Middle East in a preparation to lengthen hair and as a remedy for epilepsy. It was used in Uganda as a lightning conductor; the leaves mixed with water are said to repel soldier ants and the stem is used to filter alcohol.

Today, castor oil is used more for its properties as a lubricant rather than as a laxative but it is estimated that around 1 million tonnes of castor beans are processed every year to produce castor oil. The residue of these beans contains approximately 50,000 tonnes of ricin.

There can be problems using castor oil as a lubricant. Early aircraft, like the Sopwith Camel, used a rotary engine lubricated by castor oil. The oil would mix with the fuel/air combination in the cylinder but would only be partly burnt. The exhaust from these engines would, therefore, contain an amount of unburnt castor oil. As well as making the engines oil-hungry and, therefore, increasing the need for maintenance, this unburnt oil was the cause of dirty streaks along the fuselage. It was said that this could cause visibility problems for the pilot and so cowlings were installed to direct the exhaust flow under the fuselage. There is no mention, other than anecdotally, of the other problem, which could be expected by exposing pilots to castor oil-laden fumes.

An Australian visitor to the Poison Garden talked about a man who suffered serious diarrhoea, which cleared up quickly if he took time off work but returned when he returned to work. She was working at a poison information centre and was asked if she could suggest any possible causes. Acting on her advice it was discovered that the machine the man used at work was faulty, causing a fine mist of lubricating oil to be present in the air. The lubricant was castor oil. As soon as the machine was repaired, the man's diarrhoea cleared up for good.

In itself, excessive amounts of a laxative can be extremely dangerous, especially for anyone who is already in a poor condition. A very elderly visitor recounted his experience when a prisoner of the Japanese in Thailand during World War Two. He and about 200 other men were out on a work party when they came across a large number of bushes all bearing fruit. They fell upon them and ate the fruit, which was later identified as from the castor oil plant. All suffered severe diarrhoea and two died, though whether as a direct result of ricin or simply due to the effect of the diarrhoea can never be established.

Ricin – the undeserved reputation

But it is ricin rather than castor oil which gives Ricinus communis its reputation as a dangerous plant. In reality, however, as often happens with reputations, this one doesn't stand up to close scrutiny. It is absolutely true to say that ricin is a killer in even the tiniest amounts but those tiny amounts only rarely get to where they need to be to do harm.

Most people know the story of Georgi Markov who was murdered in 1978 by, it is believed, the Bulgarian secret service using a modified umbrella capable of firing a tiny pellet of ricin supplied by the KGB. Ricin can take several hours to produce symptoms and several days to kill but it needs to enter the bloodstream as it did with Markov. A couple of weeks before Markov's death, Vladimir Kostov, a fellow Bulgarian dissident living in Paris, was taken ill but recovered after 12 days. When the possible cause of Markov's death was becoming clear, Kostov was X-rayed and found to have a tiny pellet still embedded in his flesh. On the day he was attacked he had been wearing a thick jumper and it would appear that this slowed the pellet enough for it not to penetrate far enough for the poison to get fully into his bloodstream.

A 1985 report from the Emory University School of Medicine and the Georgia Poison Control Center examined 751 cases of accidental ricin ingestion during the previous 85 years. They found only 1.9 per cent mortality in these cases, an indication of the difficulty of causing death by ingestion of ricin. It has been suggested that with modern symptomatic support methods fatality from ingestion would be around 0.4 per cent.

So feeding ricin to someone is a most unreliable murder weapon and getting it into the bloodstream, where it would kill, is also not easy. But what about inhalation? What about all the media reports of terrorists planning to spray ricin into the air on underground railways? Though I have been unable to find out how it started, ricin is sometimes confused in the media with sarin, the gas used in the attack on the Japanese underground in 1995. Ricin is very different from sarin and would not be a practical weapon for a large-scale attack on the public.

USAMRIID – the United States Army Medical Research Institute for Infectious Diseases – publishes a 'Blue Book' about potential chemical and biological weapons. In the section on ricin, it talks about the potential effects an aerosol attack using ricin would have. In discussing treatment it says 'Superactivated charcoal is of little value for large molecules such as ricin'. It does not, however, point out that it is this large molecule size that makes ricin an unlikely choice for a large-scale aerosol attack.

What makes many people cling to the notion that ricin could be used by terrorists are the results of some work, in 1996, which showed that inhalation of ricin resulted in death in monkeys even with quite low doses. What people fail to realise is that, in this experiment, the monkeys were fitted with face masks and forced to inhale ricin. In practise, its droplets are heavy so it does not stay in the air for long enough for a significant amount to be inhaled accidentally. Presence of any sort of wind current, such as is found in underground train networks, would accelerate the fall of the ricin to the floor.

Professor John Henry, once described as Britain's best known toxicologist, died in 2007. Prof. Henry was an expert witness in the Leah Betts ecstasy case and identified the poison, dioxin, used on Victor Yushchenko, just from a photograph. According to his obituary in the *Hampstead and Highgate Express* (18 May 2007) he also:

> diffused tabloid hysteria about the planned ricin attack on the London underground by pointing out that London commuters could in fact swim in ricin without suffering any harm – it was only fatal if injected into the bloodstream.

Though ricin may theoretically be a deadly killer the sad fact is that there are plenty of better weapons available to the renegade government or terrorist wishing to mount such an attack unless, that is, intended victims could be persuaded to line up and receive an injection in the arm.

Since Markov's death, there have been 21 cases, in the USA, where ricin was a factor. Twenty of these are related to plots, threats and conspiracies. In only one is it thought ricin might have been used to attempt to commit murder but this was never established beyond doubt.

In August 1995, Dr Mike Farrar suffered four bouts of severe gastrointestinal upset, the last two of which required hospitalisation. At the time, he was in the process of separating from his wife, still called Dr Debora Green from her first marriage, prior to obtaining a divorce. On 24 September, he found packets of castor bean seeds in his wife's bag. It was later shown that these seeds had been purchased only a few days before, that is after the August attacks, and it has never been found that his wife obtained supplies earlier. The presence of ricin antibodies in Farrar's blood was discovered later indicating that his illnesses had been caused by ingesting, probably roughly crushed, castor beans.

On 23 October, Dr Debora Green set fire to the family home and killed two of her own children. Her 'no contest' plea to charges of murder, arson and attempted murder by poisoning her husband meant that a full trial was never held so the truth of the poisoning will, probably, never be known especially as Dr Green, serving a minimum of 40 years in jail, has talked of seeking to have the 'no contest' plea set aside in favour of a full trial.

It is assumed that she was trying to murder her husband but, given that he had four non-fatal attacks and that Debora Green had approximately 150 castor beans from the September purchase in her bag, it may be that she hoped to keep Mike Farrar permanently too ill to leave her. It is often said that, as a result of the ricin, Mike Farrar required heart and brain surgery. That is an oversimplification. It seems that the ferocity of the diarrhoea attacks caused his bowels to leak into his abdomen, which resulted in bacterial infections reaching his heart and brain. A number of surgical procedures were required to deal with these problems.

So, though it is 30 years, at the time of writing, since the last well documented case of murder using ricin, it remains a grave threat in many people's, and organisations', minds. An American website (Wuv'n Acres), which supplies a wide range of seeds, offers castor beans but will not supply them outside the USA and also warns:

> These seeds are sold for the sole intention of ornamental foliage in a garden setting and prevention of garden pests such as moles. We keep records of all sales for castor beans and will not hesitate to provide such records to authorized law enforcement or other professionally licensed investigative parties in the event these seeds have been used in a manner other than which they were intended, which is strictly for garden beauty.

An American visitor to the Poison Garden said that she had purchased 6 oz of castor beans online after being told that they would harm moles. A couple of days after the beans arrived she was visited by an FBI agent who knew a great deal about her and wanted to know what she was planning to do with the beans. Another time, I overheard a female visitor telling her companion how dangerous ricin was. It turned out that she was a locally employed worker at the American embassy in London where she, along with all the employees, had been briefed by the security staff at the embassy about the danger of ricin. You have to wonder if professional security staff are really so poorly informed about the true nature of ricin or whether there is a double bluff going on with the hope that 'amateur' terrorists will spend their time working on something that has no realistic chance of producing mass killings.

IS THAT CAT DEAD?

It seems improbable that local authority parks' departments know that there is no reason to be concerned about Ricinus communis being planted in municipal flower beds all round the country but people who study terror threats all the time still think it is a real hazard.

So, anyone who thinks they should avoid having laburnum or foxgloves in the garden because they think they are dangerous, though as we saw in Chapter 3, laburnum is, probably, much less harmful than its reputation suggests, needs to remember that plants in public spaces may be as full of theoretical risk as anything found in the garden. Indeed, foxgloves are often found growing wild and anyone visiting hotter climates is quite likely to find oleander growing at the side of the road.

Any advice on a plant by plant basis is almost certain to be flawed. A visitor to the Poison Garden was kind enough to send me a copy of OFSTED's 2001 publication 'Childminding: Guidance to the National Standards', which has an appendix listing 'Poisonous Plants'. The guidance says childminders should consider how they will supervise children in the vicinity of poison plants, how children with allergies might be affected and whether to prevent access to areas that have these plants. The problem is that, as OFSTED itself states, the list is incomplete. It does not, to take two examples, include bryony, which, being a plant that produces berries, might be attractive to children or daffodils, which might be found in the garden shed or garage.

Other publications on poison plants often have a degree of 'crying wolf'. One American book, intended as a guide for parents, says holly should not be brought into the house at Christmas because the berries are poisonous, but the Kew Gardens' guide to poison plants gives only one case report where a two-year-old suffered two hours of vomiting after ingesting two berries. In my opinion, it is impossible to stop children having the opportunity to ingest poison plants so what matters is to educate them, from an early age, not to eat anything that a parent or grandparent has not seen and approved. It is particularly important to educate them not to be coerced into trying something by a friend who claims that it is harmless or calls them names for declining. Inculcating that piece of advice will pay dividends later in life when they will be encouraged to try drugs by some of their peers.

The overwhelming majority of plants add beauty to our world and it would be foolish to deprive ourselves of the pleasure of seeing them because of an exaggerated concern over their ability to do harm. As far as the United Kingdom is concerned, there is only one plant that should not be allowed to grow in the garden and that it not because of its poisonous potential. Indeed, many people would argue that is is not a poison plant at all. That being so, just why was cannabis a central part of the Alnwick Garden Poison Garden?

13

Where's the Cannabis?

Quite a number of people expressed the strong opinion that cannabis should not be included in a poison garden because it is not a poison. It is true to say that cannabis is not toxic in the sense that there is no such thing as a fatal dose but there were plenty of other plants in the Poison Garden, like Ruta graveolens, rue, or Polygonatum odoratum, Solomon's seal, which have no history of causing death by ingestion. If the definition of poison is taken as 'capable of causing harm' then cannabis is every bit as much a poison plant as any other you might mention.

Cannabis sativa is a big draw in the Poison Garden. Many people, especially pre-teens and early teens, only have one question on their minds when they get to the Poison Garden gates: 'Where's the cannabis?'

Like the mandrake of Chapter 10, whole books have been written about the cannabis plant and with many news stories every day about this plant and its use and abuse enough words for a whole book are being added to the total every few days. The intention of this chapter is to look at some of the history of this plant, how deeply embedded in our society it has become and just some of the issues surrounding its use today and the effects of its illegal status.

History and herbals

There is no clear indication of when cannabis first came to the attention of the human race, though 10,000 years ago is often given as an approximate figure, but the stories of how its name arose give some indications of its long history. There are many alleged origins of the word 'cannabis'. It is said to come from the Assyrian 'qunubu' meaning 'way to produce smoke', but Martin Booth (2005), in *Cannabis, a History*, says the Assyrian word was 'qunnabu', meaning 'noise'. Many ancient languages, however, use some form of the word 'kan' to mean 'cane'. The Hebrew 'bosm' and the Aramaic 'busma' both mean 'aromatic' and may have led to the third syllable 'bis'. Booth concludes that 'cannabis' is the 'fragrant cane'. This derivation seems to accord with the view that the kaneh-bosm mentioned in Exodus 30: 22–25 is cannabis and not camalus as in most translations since early times. Camalus is Acorus camalus or sweet flag, a marsh plant of little value but, in Ezekiel 27:19, it is said that goods have been bought by exchange for 'wrought iron, cassia and kaneh', suggesting that kaneh-bosm is valued.

Many of those who campaign for the legalisation of cannabis suggest that the current prohibition is against the run of history because cannabis has always been a useful plant to the human race with a wide range of medicinal uses. In older cultures, medicinally, cannabis has been used as an antispasmodic, analgesic, tonic, intoxicant, stomachic, narcotic, sedative and anodyne. Seeds and leaves were used to treat old cancer and scirrhous tumours. The seed, either as a paste or as an unguent, was supposed to be a remedy for tumours and cancerous ulcers. The decoction of the root, it is claimed, helps remedy hard tumours and knots in the joints.

The leaf, prepared in various manners, supposedly alleviates cancerous sores and tumours. Cannabis is listed as having folk medicine uses in a huge number of conditions: alcohol withdrawal, anthrax, asthma, blood poisoning, bronchitis, burns, catarrh, childbirth, convulsions, coughs, cystitis, delirium, depression, diarrhoea, dysentery, dysmenorrhoea, epilepsy, fever, gonorrhoea, gout, inflammation, insomnia, jaundice, lockjaw, malaria, mania, mennorrhagia, migraine, morphine withdrawal, neuralgia, palsy, prolapse of the uterus, rheumatism, scalds, snake bite, swellings, tetanus, toothache and whooping cough.

But, often, when looking at the history of a plant it is important to look at what is not said. Pliny the Elder says that hemp seeds make the genitals impotent. The juice of the seeds drives out worms and other creatures from the ears but gives a headache. The root, boiled, eases gout, cramp and other 'violent pains'. John Gerard (1598) gives only a brief description of its 'Vertues' and in these draws heavily on what earlier writers had to say rather than citing contemporaneous uses. His full description reads:

> The seed of Hempe, as Galen writeth in his bookes of the faculties of simple medicines, is hard of digestion hurtfull to the stomacke and head, and containeth in it an ill juice: notwithstanding some du use it to eate the same parched, cum alus tragematis, with other junkets. It consumeth winde, as the said Author saith in his booke of the faculties of medicines, and is so great a drier, as that it drieth up the seed if too much be eaten of it.
>
> Dioscorides saith, That the juice of the herbe dropped into the eares asswageth the paine thereof proceeding (as I take it) of obstruction or stopping, as Galen addeth.
>
> The inner substance or pulpe of the seed pressed out in some kind of liquor, is given to those that have the yellow jaundice, when the disease first appears, and oftentimes with good successe, if the disease comes of obstruction without an ague; for it openeth the passage of the gall, and disperseth and concocteth the choler through the whole body.
>
> Matthiolus saith, that the seed given to hens causeth them to lay eggs more plentifully.

William Turner doesn't even mention cannabis in his three-part herbal and Culpepper repeats Gerard's remedies but, interestingly, notes that hemp is such a well-known plant that it doesn't require description. You have to ask, well-known for what?

On the ropes – hemp fibres

Those who would have you believe that cannabis has been used by the human race to get high throughout its long history, nonetheless, usually cite its value as a provider of fibre when giving details of its past worth. Perhaps, this use explains why Culpepper was so sure his readers would be familiar with the plant. It must have been of some great value given that Culpepper, whose herbal was published in 1653, was so sure of its familiarity only 34 years after it was first discovered in England. In 1629, Thomas Johnson, who went on to extensively revise Gerard's *Great Herbal* for its re-publication in 1633, in what is now accepted as its definitive version, joined an expedition of apothecaries intent on exploring an area and documenting, for the first time, the plants found there. The expedition was to Kent. One of the plants they brought back, which had never before been found to grow in England, was Cannabis sativa.

One indication of just how useful cannabis fibre was can be seen from a word which it gave the English language. Frans Hals' painting, which has been known since Victorian times as *The Laughing Cavalier*, was painted in 1624. The answer to the question 'Why is the cavalier laughing?' is simple: 'because he is on cannabis'. The material on which the painting is executed is canvas – a word that derives from the Dutch word *canefas*, meaning 'made of cannabis'. Until its replacement by cotton or linen, cannabis fibre was an essential part of sailing ships not just for the canvas sails but also for ropes.

Speaking of hemp ropes brings me to a little aside, which is illustrative of how people misinterpret history. It is sometimes said that the hemp rope used on ships was coated in tar to stop sailors chewing it. The hemp rope on ships was tarred at the ends to prevent fraying and the pine pitch that was used (before bitumen) to do this was chewed by sailors as a sort of chewing gum. So using this to suggest that sailors thought they could get high by chewing hemp ropes is simply untrue.

WHERE'S THE CANNABIS?

Talking of the words cannabis has given to the English language, 'roach', as many people know, is the name given to the butt end of a cannabis joint. The word comes from Mexico early in the 20th century. The song 'La Cucaracha' appears to have originated in Spain, possibly as early as 1492, and to have had many different verses written to suit different circumstances. 'La cucaracha' is 'the cockroach' and the term seems to have been applied to foot soldiers from many armies and bandit gangs. In Mexico, Pancho Villa's men were referred to as cockroaches and the verse appropriate to the time goes:

> La cucaracha, la cucaracha
> Ya no puede caminar
> Porque no tiene, porque le falta
> Marihuana que fumar.

Which translates as;

> The cockroach, the cockroach
> Now he can't go travelling
> Because he doesn't have, because he lacks
> Marijuana to smoke.

Hildegard and hashish

Saying that cannabis was most valued for its fibre is not to say that its psychoactive effects were unknown. Tradition in India maintains that the gods sent man the hemp plant so that he might attain delight, courage and have heightened sexual desires. In *De historia stirpium* (1542), Fuchs adds his own views on cannabis:

> Since it is clear ... that cannabis excites the mind and so injures it, those who unwisely, following a common error,

administer potions of this seed to those with mental ailments, especially the serious ones, do so with great risk to the patients.

Hildegard of Bingen was a 12th-century German mystic. About hemp she says:

> ...whoever is weak in the head and has a vacant mind, if that person will have eaten hemp, it easily makes the person suffer pain somewhat in his or her head. However, whoever is sound in the head and has a full mind, it does no harm.
>
> (von Bingen, 2007)

So, the idea that cannabis could be helpful to some and harmful to others is not a new one. Nearly 900 years after Hildegard made this comment, based purely on her observations of its effects, we are not much closer to understanding how to identify those it will harm and those who seem to suffer no ill-effects. One of the obstacles to reaching a conclusion on this is that we still do not really know what we mean when we say 'cannabis'.

To look at this further we need to define a few terms. Cannabis has many, many common names applied to it and, as much as some sources attempt to give dogmatic definitions of the main ones, these names do get shifted around by both those ignorant of cannabis culture and its adherents. In general, it can be said that 'marijuana' is the name given to the pressed flowering tops and 'hashish' is the sticky resin that forms around the buds. Cannabis is a dioecious wind-pollinated plant; that is to say the male and female are different plants and pollen from the male blows onto the female. Though no one has been able to definitively state the purpose of this resin it seems possible that its stickiness helps it to catch the male pollen blowing by on the wind. Certainly, it is known that in the absence of a male plant the female will produce more and more resin, which does suggest it is trying to increase its chances of catching the male pollen. The effect of increasing resin production will be examined later in this chapter. 'Hash oil' is the oil that can be pressed from the seeds and has the highest concentration of the main psychoactive ingredients.

All substances come in a variety of forms and, hence, have a variety of effects, but this is truest when it comes to cannabis. With marijuana (the pressed flowering tops), hashish (the resin extracted from around the buds), hash oil (the essential oil) and, for some, consumption of the leaf and stalks, the forms in which cannabis can be consumed offer a wide range of active ingredients. Add to that differences in the means of consumption, smoking a joint, pulling a bucket, using a vaporiser, putting in food, etc., and the possible range of doses increases considerably.

Then, with cannabis, there is the added dimension of its complexity. There is increasing evidence that the balance between THC (delta–9-tetrahydrocannabinol), CBD (cannabidiol) and CBN (cannabinol), the three main active ingredients, plays a major role in the effects of cannabis. But, overall, there are some 500 compounds in cannabis of which about 70 are believed to be psychoactive.

Strength, sinsemilla and silicosis

Making a universal statement about cannabis whilst ignoring all the possible differences in potency is ridiculous in itself but becomes more so when the question of contamination and adulteration are taken into account. There are people who claim that all the reported adverse effects of cannabis are due to contaminants and nothing to do with the 'pure' substance. That claim may be taking things too far but there was recently research on the content of the vapour given off by the electric vaporisers that are touted as ways to safely inhale cannabis without the tar produced by smoking. This work found high levels of ammonia in the fumes generated from all the samples of 'street' cannabis tested but comparison tests with 'standardised' cannabis produced by the US government for scientific research did not exhibit these high levels. Whether the ammonia arose from the higher strength of street cannabis or as the result of contamination is impossible to say.

The July 2008 edition of *Addiction, the Journal of the Society for the Study of Addiction*, included a review of the literature on cannabis potency and contamination. The review found that there has been no systematic measurement of cannabis potency over time. With wide variations in strength within any period, it is impossible to state with certainty that cannabis has increased in strength in recent years. They also found that there has been no systematic monitoring of contamination when studying the effects of cannabis.

The authors conclude that:

> Claims made in the public domain about a 20- or 30-fold increase in cannabis potency and about the adverse mental health effects of cannabis contamination are not supported currently by the evidence.
>
> (Allsopp, Dillon, McLaren, Swift, 2008)

These claims about higher strength cannabis are based on the increasing tendency of growers to produce sinsemilla plants. 'Sinsemilla' is from the Spanish for 'without seed' and refers to a female cannabis plant that has not been exposed to a male and is, thus, unfertilised. As mentioned earlier, in this situation the female produces more of the sticky resin around its buds. It is not completely clear whether the resin increases in strength as well as amount or whether the higher proportion of resin in a sample of pressed buds (what is usually called 'marijuana') means that the average strength of the marijuana is higher. It is often said that sinsemilla is a recent development but this way of growing cannabis was described at least one thousand years ago in India though the people who discovered it are not recorded as having used cannabis for anything other than fibre.

WHERE'S THE CANNABIS?

What has changed is the knowledge of this growing technique and the increase in indoor cultivation making it possible to keep pollen away from the female plants. The ease with which a female plant can be pollinated is shown by our experience in the Alnwick Garden Poison Garden in 2007. Cannabis for the Poison Garden was grown from seed and the gardeners would weed out and destroy all the male plants, as soon as they could be identified, as female plants produce a much better show. In late May 2007 a sturdy unfertilised female plant was brought out from the greenhouse and planted in one of the cages described in Chapter 14. Within a few weeks it was obvious that the plant was not developing as the 2006 plant had and closer examination revealed that the plant had been fertilised and was producing seeds. The conclusion drawn from this was that someone, quite close to the Alnwick Garden, must have a male plant growing in their garden. Since those who grow cannabis to indulge in it understand its botany very well, it is almost certain that this was a plant that had sprung up from birdseed.

'It must have grown from birdseed, officer' sounds like one of those excuses that criminals hope will get them let off without charge but, in fact, cannabis can grow from birdseed and other products as several visitors to the Poison Garden could attest. Hemp seed intended for sale has to be sterilised by heat but there is only a narrow range between the temperature that will 'kill' the seed and that at which the seed will split, leading to rotting, which can spread to a complete batch. This, inevitably, means that some seeds in any batch being heat-treated will not reach the required temperature and remain viable.

Hemp seed is not, usually, included in wild bird food so it is only if caged bird food drops in the garden that germination may occur. Cannabis plants growing innocently in the garden are, thus, fairly rare and usually only found as one or two plants. Except for a female visitor who explained how she mistook her husband's fishing bait for garden fertiliser and spread it around her plants. Some, if not all, of the bait had missed the heat treatment process intended to make it sterile and she was soon growing a healthy field of cannabis. It was her postman who identified it and, after that, she could never see the postman without wondering how he knew.

The key question with cannabis is, of course, is it harmful? The reality is that it is very difficult to know, both because of the issue of not knowing what strength is being examined and also because of the question of contamination and adulteration.

Contamination is the accidental addition of other substances to cannabis in the course of its preparation. It is sometimes suggested that other substances like strychnine are added to cannabis to increase the high. Though there appears to have been no detailed study, it does seem that cheaper grades of cannabis resin may well be contaminated with a variety of unpleasant substances as a result of poor handling techniques rather than deliberate adulteration.

There is, also, another possible explanation for these stories of adulteration. Robert Connell Clarke, in *Hashish!*, explains that hand-rubbed cannabis resin may contain more moisture than is desirable and may also be contaminated by the simple fact of its being produced by hand-rubbing. This can lead to the formation of mould, often in the form of white threads, which becomes visible when a block of resin is broken open. Dealers, faced with the prospect of having to discard stock that has degraded in this way are thought to tell gullible buyers that the threads are composed of opium or other substances which have been deliberately added to increase the effect of the cannabis.

There have also been reports of contamination with diesel oil, which, for some reason, has been used in the preparation process. It should be no surprise that those who are preparing an illegal product for sale are unlikely to be too concerned about the quality of that product and may well take short cuts to speed its production and, thus, accelerate their chance to cash in.

This unrestrained desire for profit shown by illegal growers also leads to adulteration, which is the deliberate adding of foreign substances intended to either increase the bulk of the product or to increase its apparent strength so that it will command a higher market price.

One of the most worrying stories about adulterated cannabis concerns so-called 'grit weed', which may well be incubating health problems for young people. First reports of 'grit weed' began appearing on the Internet in February 2006 but it was only in 2007 that it became more widely available. It would seem that the clamp-down on cannabis production towards the end of 2006 may have made the situation worse because a reduction in availability led regular users to turn to different suppliers with whom they did not have a relationship based on trust.

Smoking 'grit weed' can lead to coughing and sore throats and may, in the longer term, result in silicosis, a potentially fatal lung condition. Whilst there is evidence of cannabis being contaminated with a variety of substances to bulk it up and increase the dealers' profits, it would appear the true 'grit weed' is a more sophisticated scam.

To produce 'grit weed', growers are spraying plants with an industrial spray used to produce frosted glass about seven to ten days before harvesting. This deposits small glass beads on the buds of the plant and allows time for further cannabis resin to form around them. When harvested, the beads are fully incorporated in the resin and can glint in light in a similar way to the glint produced by crystals of THC, the main active ingredient in cannabis. Thus, 'grit weed' looks, on simple examination, to be high THC resin.

So, as well as being bulkier which means that an ounce of resin may appear to be up to 25 per cent heavier, true 'grit weed' can be sold at a higher price per ounce because it appears to be higher quality. It is said to have originated in the Netherlands and is only sold to the UK as only UK users are 'stupid' enough to accept it. Other reports, however, suggest that it is produced in two Welsh cities by two well-known gangs.

Smoking 'grit weed' has already caused problems of coughing

and throat irritation but it is the longer term effect that is the more worrying. Silicosis is a severe lung disease caused by inhalation of silica particles. It caused many deaths amongst mine-workers and stone-cutters as many rocks contain silica particles and silicates. Speaking on ITV Wales in 2007 Eggsy, a member of the rap group 'Goldie Lookin Chain', said that two puffs of a joint made with 'grit weed' had given him a sore throat and a friend of his had lost his voice for a time.

Many stories about adulterated drugs that appear in the mainstream press turn out to be distortions or exaggerations and present a picture that is far scarier than the reality. In the case of 'grit weed', however, it seems that its potential harm cannot be overstated. On Internet blogs associated with the use of cannabis and other drugs there are numerous postings warning of the dangers and pleading with users to be on guard against being offered 'grit weed'. There are also suggestions that some people are turning to other substances instead of cannabis without considering the harm those substances do.

There are some voices that say that, as cannabis use is illegal, any harm done to users by 'grit weed' 'serves them right'. My attitude is that abusing any substance that risks damaging your health is a foolish thing to do and it doesn't matter whether the substance is legal or not. The production and sale of 'grit weed' is just another example of how unscrupulous suppliers are willing to exploit their customers for the sake of increased profit.

The dose dilemma

Aside from the harm which might result from using contaminated cannabis, there is also the question of the dose being inhaled and whether this has increased in recent years. There is no doubt that new methods of smoking cannabis do have the potential to greatly increase the amount inhaled on each occasion. The most potent of these is known as 'pulling a bucket'.

WHERE'S THE CANNABIS?

Cannabis burns at a much higher temperature than tobacco and one limit on dosage has been the ability to tolerate this high temperature when inhaling from a joint. The 'bucket bong' was developed as a way of cooling the smoke before inhalation. It involves the use of a plastic bottle, a bucket of water and a 'mix cone', a metal ring designed to screw onto the top of the bottle. This has a cone-shaped hole through it so that a 'mix' of cannabis will sit in the cone without falling through but air will pass through the mix to facilitate burning. The following description comes from a website that actively promotes the use of cannabis.

> The bucket bong is another full-hit wonder. It's simple to prepare. Get your big bottle and cut away the bottom. Put the bottle in the bucket so it touches the bottom. Then fill the bucket with water. The water level should not rise above the bottle neck. Place the 'mix cone' onto the mouth piece of the bottle and apply flame to the 'mix' continuously as you raise the bottle away from the water. This creates suction and you'll see the smoke being drawn into the bottle-chamber.
>
> Once the bottom of the bottle has just reached the waters surface, quickly remove the 'mix cone' and place your mouth over the bottle's mouth piece, then push the bottle down again, while you inhale the smoke getting pushed up.

The 'mix cone' used to hold the 'mix' at the top of a bucket bong is usually a small metal bowl threaded at the bottom to fit securely into the bottle or whatever is being used as a 'bong'. This cone contains enough cannabis for one 'hit'. You can, however, buy a 'Six Shooter' mix cone, which is made up of a ring of six cones enabling six hits to be taken in rapid succession. The promotional material for the 'Six Shooter' says 'So ya think you can take your bongs do you?' It then describes how to use the device and concludes: 'Think you can handle it?'

On the face of it, someone using a bucket bong to smoke cannabis is going to absorb a large amount of the drug with each puff but, in reality, that is not the case. Different users have varying techniques for smoking from a bucket bong. I've seen people remove the bottle from the water and inhale the smoke as they would if smoking a joint. I've seen people leave the bottle in the water and inhale the smoke by drawing hard against the water, effectively like trying to suck all the air out of an intact bottle. And I've seen people doing what may be considered to be the 'classic' bucket bong technique of driving the bottle down into the water to force the smoke out. But, even here, there are variations. Some people, having inhaled a large volume of smoke, will hold it in their lungs for as long as possible to maximise the transfer of the active ingredients and, when they exhale, there is very little smoke. Others exhale as soon as they finish inhaling and, therefore, get a much smaller dose from what appears to be the same technique.

This problem of knowing the precise amount of cannabis an individual user is consuming makes drawing firm conclusions from research very difficult. There are many reports on the effects of cannabis but just two of them illustrate the dilemma of not knowing exact amounts.

The first, published in the August 2007 edition of *Addiction*, de Graaf, Monshouwer, van Dorsselaer and van Laar (2007), reported the analysis of a Dutch mental health survey conducted between 1996 and 1999. This seemed to show that there was an increased risk of depression or bipolar disorder but not anxiety disorders for cannabis users. The authors state that their results may be imprecise as the average age of the survey participants in 1996 was 39. Most people likely to exhibit mental illness will have done so before this age and anyone with existing mental health problems was excluded from the study. It would be wrong, therefore, to assume that the increased risk of illness apparently found can be taken as demonstrating that cannabis is solely responsible for it.

The second report, published in the July 2007 edition of *The Lancet*, Barnes, Burke, Jones, Lingford-Hughes, Moore and Zammit (2007), was a so-called meta analysis of other studies. That is to say, the authors looked at a number of studies and tried to draw them together to look for overall results. This type of analysis can be problematic as the authors have to make assumptions about the interchangeability of data from studies conducted in completely different ways. This study appeared to show that any use of cannabis, even once in a lifetime, gives a 40 per cent increase in the risk of psychotic illness in later life. One of the authors, however, interviewed for *The Lancet* website said that it was not possible to say that there was a definite, a so-called causal, link between cannabis and psychosis. It is possible that people who are willing to break the law to use cannabis have some difference in their make-up which makes them more liable to psychosis. Also, use of any other drugs was not assessed for any of the studies so the chance that other substances may have caused the psychosis must not be ignored.

In both cases, the studies relied on self-reported use of cannabis, which removes any chance of estimating any data relative to dose.

Shedding light or muddying the water?

The question of a link between cannabis and schizophrenia surfaces from time to time, often following a murder. Two cases illustrate the difficulty of putting the blame on cannabis, alone. In the first case, where a young man killed two friends while they were out in the country, as well as smoking cannabis, the three had been drinking heavily. In the second, where a man stabbed to death a complete stranger after hearing voices telling him to find someone to kill, the killer had been a regular user of cocaine as well as cannabis and had lost any sense of reality after constantly playing the computer game Grand Theft Auto. In the second case, the killer was not under the influence of cannabis at the time of the murder.

In an attempt to finally determine whether cannabis causes schizophrenia, a paper in the April 2007 issue of *Addiction*

(Hickman, Jones, Kirkbride, Macleod and Vickerman, 2007) *assumes* there is a link and projects the number of cases of schizophrenia that will be seen in 2010. When 2010 comes it will be possible to see if the actual number of cases is anywhere close to the estimate which should indicate whether there is a causal link between cannabis and the disease.

There are even some research reports suggesting cannabis may be helpful. Analysis of a survey of Swiss adolescents, conducted in 2002 (Akre, Berchtold, Jeannin, Michaud and Suris (2007)), suggests that young people who use cannabis achieve a higher educational standard than those who also use tobacco. At the same time, the 'cannabis only' users play more sport and get drunk less often. Cannabis users also play more sport than those who use neither tobacco nor cannabis, but have worse parental relationships. The authors conclude that, whilst cannabis use of any sort should not be trivialised, more focus should be placed on the needs of tobacco-smoking cannabis users.

In a paper published in *Scientific American* (Arnold, 1989), researchers at the University of California, Los Angeles, reported finding no link between smoking cannabis and an increased risk of lung cancer. The study, involving over 2,000 people, matched cannabis users and non-users, both cancer-free and lung cancer patients, and found that the cannabis users showed no increased incidence of lung cancer even for long-term very high consumption users.

Prohibition

So, if the evidence of cannabis causing harm is incomplete, how does it come about that cannabis is an illegal substance? Many people think the first attempts to outlaw cannabis were in the USA in the 1930s but other countries had looked at the issue in the 19th century.

In Egypt, for example, action against cannabis began in 1868 but with little effect. In 1884, to try and motivate local officials to enforce the law they were allowed to sell seized hashish overseas and use the money raised to pay informants and themselves. Lacking overseas contacts, they sold to dealers who were able to sell the hashish overseas. In many cases, however, it was easier, and more profitable, for the dealers to sell the drug back into the local market.

In 1917, Thomas Wentworth Russell, known as Russell Pasha, was appointed head of the Egyptian Central Narcotics Intelligence Bureau. Russell was unsure about the merits of suppressing cannabis use, preferring to concentrate on cocaine and opium and its derivatives. This was, in part, because he had evidence that if cannabis users were unable to obtain the drug they would add henbane leaves to tobacco.

But it was the 1930s in the USA that saw the start of the present official attitude towards cannabis. Henry J. Anslinger was appointed as the first Commissioner of the Treasury Department's Federal Bureau of Narcotics on 12 August 1930. Some say Anslinger was a natural prohibitionist who wanted to find something to ban soon after the end of the Prohibition era. Others say he was concerned at the number of federal agents who would become unemployed with the legalisation of alcohol though it is also said that he saw this pool of available agents as a way to build a large empire for himself very quickly. The most extreme allegation against him is that he was in the pay of Randolph Hearst whose investment in pulp mills for making paper for his newspapers was at risk if cannabis fibre was able to offer an alternative source of paper.

Whatever the true reason, Anslinger set out to 'get' cannabis and there is clear evidence that he regularly lied about marijuana's effects. It is also true that Hearst's newspapers fully supported his campaign and the alliterative phrase 'marijuana madness' frequently appeared in headlines. Anslinger's biggest hurdle was that plants that could or could not be grown was a state, not a federal matter so, if an individual state decided that hemp fibre would make a useful contribution to its agricultural income Anslinger could not ban it from allowing cultivation.

IS THAT CAT DEAD?

It took until 1937 before Anslinger had so swung public opinion against cannabis (to the point that people who had never come across it would say, with certainty, that it was a dangerous menace) that he was able to get the federal government to act within its authority by passing the Marijuana Tax Act. The federal government didn't have the right to outlaw the growing of cannabis but it did have the right to raise taxes so the Marijuana Tax Act introduced taxes on the growing of cannabis, which were so high that it became impractical for anyone to grow it. Later, the federal government would use its authority to ensure international treaties were observed as its way to control narcotics.

One of Anslinger's arguments was that cannabis produced what he called amotivational syndrome. That is to say users became indolent and lazy and were a drain on society because they would do no work. The reality is rather different. As noted above, the traditional song 'La Cucaracha' says that Pancho Villa's men couldn't go travelling unless they had marijuana to smoke, i.e. the marijuana enabled them to put in a greater effort.

For two years (1970–72) American researchers studied cannabis use amongst working-class Jamaicans. The project included very careful measurement and analysis of agricultural work done both before and after smoking cannabis. A key finding was that workers expended much more energy after smoking and would take much greater care when digging over and weeding a patch of ground. It should be said that there is no indication that the additional work produced any real benefit and may have been a genuine waste of effort but, in the context of the alleged 'amotivational syndrome' the willingness to work is what is of interest. The authors conclude:

In all Jamaican settings observed, the workers are *motivated* to carry out difficult tasks with no decrease in heavy physical exertion, and their perception of increased output is a significant factor in bolstering *their motivation to work*. [Emphasis added.]

(Comitas and Rubin, 1975)

WHERE'S THE CANNABIS?

Many of Anslinger's arguments against cannabis continue to be used today to justify its prohibition in spite of there being no evidence to support them, and evidence suggesting that cannabis is not especially problematic continues to be ignored. Even the United Nations Office for Drugs and Crime in its 2006 'World Drug Report' notes that regular, high users of cannabis often limit their use to their own leisure time and do not use it during work hours so it is clearly not in the same league as heroin or alcohol where high users become addicted to the point that they must have their substance whatever the time or circumstances.

Official estimates show that some 29,600,000 Americans in the age range 15 to 64 use cannabis at least once a year. Figures from the CIA estimate the US population in that age range as 200,400,000. This means that 14.8 per cent, or one in seven, adult Americans uses cannabis. For England and Wales the comparable figure is 10.8 per cent while Scotland is lower at 7.9 per cent. Overall, the 2006 report by the United Nations Office for Drugs and Crime estimated that about 1 in 25 of the world's population in the age range 15 to 64 uses cannabis at least once a year.

The figures show that the number of 'problem' cannabis users is only a tiny percentage of the whole, which, qualitatively, suggests that cannabis cannot be as harmful as governments would like us to believe, though it must be accepted that cannabis is harmful to some of those people who use it.

What can not be known is how much of that harm is due to the unregulated strength of cannabis sold on the street and how much is due to contamination or adulteration rather than to some inherently harmful effect of pure cannabis. What, to me, seems clear, however, is that, accepting that cannabis can be harmful, it is morally unjustifiable for governments to leave its manufacture and distribution in the hands of criminal gangs whose only motivation is quick profit.

IS THAT CAT DEAD?

Tea, weed and Mary Jane

Cannabis not only has a great many common names, some of which it shares with other substances, it has also given other words and phrases to the language. I'll finish this chapter by looking at four of the terms associated with cannabis.

Particularly in 1920s New York, marijuana was known as 'tea' and was available from so-called 'tea pads'. It is believed that the first of these was set up following the passing of the 18th Amendment and the Volstead Act of 1920, which introduced the prohibition of alcohol, but it was the late twenties before the term became widely known. It is said that the song 'Tea for Two', from the 1925 musical *No, No, Nanette*, was an insider joke for those who were in touch with Harlem society enough to realise that what was being suggested was not 'the cup that cheers but not inebriates'.

While looking into the origins of the term 'tea pad' I came across a 1944 report into marijuana in New York, which gives an interesting insight into the assumptions made about it. It offers this description of a 'tea pad':

> Usually, each 'tea-pad' has comfortable furniture, a radio, victrola or, as in most instances, a rented nickelodeon. The lighting is more or less uniformly dim, with blue predominating. An incense is considered part of the furnishings. The walls are frequently decorated with pictures of nude subjects suggestive of perverted sexual practices. The furnishings, as described, are believed to be essential as a setting for those participating in smoking marihuana.
>
> (La Guardia, 1944)

I imagine many art lovers would dispute the alleged link between depictions of the nude and 'perverted sexual practices'.

WHERE'S THE CANNABIS?

As I have already mentioned, cannabis burns at a higher temperature than tobacco. When smoking became known in India the first attempts to consume cannabis in this way were unsuccessful until someone modified a pipe to have a bowl of cannabis, which would increase the time between the smoke being created and inhaled so as to allow some cooling. This pipe, after the Hindi word for bowl, 'chilum', became known as a 'chillum'. When western young people (from both Europe and America) hit the 'hippy trail' in the 1960s they soon adopted the term 'chill out' as an invitation to consume cannabis.

Not that the hippies were the first Americans to discover cannabis. The substance crossed the Mexican border into the USA early in the 20th century. In Mexico, cannabis was, primarily, available from brothels. When Mexican men were planning a visit to a brothel they would say they were going to see Mary and Jane, in Spanish, Maria y Juana. This name for prostitutes became attached to cannabis and has remained with it ever since.

Cannabis is also frequently called 'weed', probably for two reasons. It is a very vigorous plant, growing to over two metres in one season and any gardener will say, usually with a sigh, that weeds are the plants that grow best. Secondly, there are parts of the world where it does grow in the wild as a weed. One visitor to the Poison Garden remarked that he had made a trip to South Africa to see a friend who lived there. His friend took him out for a ride in the bush and they came round a bend to be confronted by the sight of cannabis plants growing freely. The visitor was, in his own words, 'nearly struck dumb' but his friend was very casual about it, saying that it grows in the wild and, hence, is called 'weed'. Whilst not disputing this, our visitor said it was the first time in his life that he had seen wild plants growing in perfectly straight rows.

14

Why Are Some Plants in Cages?

Actually, this was not a question that many visitors asked because they tended to provide their own answer before getting around to posing the question. The layout of the Poison Garden was a long narrow space with flame-shaped beds surrounded by paths for visitors to walk about or stand in listening to a tour. In total, there were six black metal cages in various of the beds. Many people would see the cages and say, 'I assume those are the most poisonous plants, which is why they are in cages' but that was not the basis on which the plants were selected.

The provision of cages, in the first place, was because, during the protracted negotiations with the Home Office before the granting of the licence to have Erythroxylum coca and Cannabis sativa, the question of how to keep the public away from these plants was a concern. Having come up with the idea of metal cages so that visitors could look but not touch, it was decided to provide additional cages so as to complete the look of the garden and provide a third dimension, especially during the winter when most plants would not be growing to any great height.

The decisions about which plants to put in cages were all made before I became involved with the project so this chapter deals not with the reasons these plants were put in the cages but the reasons I gave to visitors for their selection.

Of the six cages, two were to keep the public away from the plants which were covered by the Home Office licence (and both Erythroxylum coca and Cannabis sativa have been discussed at length already). Quite what the Home Office thought would happen if visitors were able to touch, and even, perhaps, remove a leaf or two from these plants I can't say, but there they were in their cages, whether that was really necessary or not. The third cage was absolutely essential, because the plant in it, Heracleum mantegazzianum, the giant hogweed, is capable of doing serious harm purely from contact, and the three others drew attention to plants with interesting stories.

Giant hogweed – unwelcome, unpleasant and under attack

Though many visitors had heard of the giant hogweed, many had not and were surprised to find that such an unpleasant plant exists. The plant, which is capable of growing to anything up to 5 m or more, contains furocoumarins (psoralens), which produce changes in the cell structure of the skin reducing its protection against the effects of UV radiation, and exposure to sunlight after brushing against the plant can cause severe skin rashes and/or blistering and burns. These furocoumarins are present throughout the plant and many people have suffered burns just from brushing through a clump of giant hogweed but they seem to be most concentrated in the juice in the stems so crushing a stem so as to release the juice can cause a severe reaction.

Some of the most unpleasant cases, which visitors passed on, involved strimming a piece of rough ground, early in the summer, before the giant hogweed had grown above the rest of the vegetation. The strimmer would break the stems into small pieces, which would fly through the air. If the person doing the strimming was wearing a short-sleeve shirt, or no shirt, they would find small red patches appearing, usually the next day. These might cover a considerable part of the torso and produce much discomfort. But the worst part about giant hogweed is that, having removed the UV protection from the skin, the skin remains sensitive to light for two to three years afterwards. We had reports of people who,

WHY ARE SOME PLANTS IN CAGES?

forgetting their previous exposure and the warnings they received at the time, had removed their shirts on the first hot day of the following summer, or the second summer after, only to find themselves burning again.

Giant hogweed was introduced to the UK from Russia by the Victorians who thought its size would make a dramatic statement in large gardens. It thrived, having left its natural enemies behind it and then escaped and has spread rapidly to be a major problem on river banks in some areas. The Wildlife and Countryside Act 1981 makes it an offence to plant or cause giant hogweed to grow in the wild.

Each giant hogweed plant is capable of producing about 50,000 seeds and, though they only drop close to the plant, they can be transported on shoe soles, or carried along by the rivers, which the plant likes to be beside, to other areas. The seeds remain viable for seven years, meaning that eradication is a long and expensive process. After spending £250,000 in two years it was claimed that there was no flowering giant hogweed on the River Tweed in 2005. Since then, annual spraying in the spring has been undertaken to keep the plant down, but in June 2008 I saw a plant in full flower on the banks of a tributary of the River Tweed.

The only alternative to a seven-year programme of spraying is to completely remove the topsoil, which may contain the seeds. One visitor described how he had used a specialist contractor to undertake complete removal with a guarantee that it would not return at a cost £20,000 for an area, which he said was not much bigger than a small town garden.

Obviously, the prospect of any visitor to the Poison Garden having contact with such an unpleasant plant could not be contemplated and so the plant was kept in a cage. There was a bit of a problem in that the cage was only about 2 m high so the giant hogweed would completely fill the cage. In 2006, three plants had been put in and it became necessary to cut them back. It took six gardeners wearing full head-to-toe suits and looking like something out of a biological warfare film to complete the task: four to remove the cage and two to cut the plants back. In later years only one plant was placed in the centre of the cage.

That exercise gave an excellent demonstration of the idea that plants want to grow and reproduce. In order to get the cage back over the plant, it was necessary for the gardeners to cut it almost down to the ground. The plants had been at the flowering stage but had not set seed. Within about three weeks of this drastic attack on them, they put on new stems and, when these were only about 1 m high, produced flowers. Again, the flowers were removed to prevent the seed setting and one of the plants put up a very short stem, less than half a metre high, and flowered for a third time.

So, two cages for the plants that needed a Home Office licence and one for the plant that could not be touched. That left three cages where the plants were selected to draw particular attention to some aspect of their story.

In Chapters 4 and 6, I explained how the Datura/Brugmansia plants could kill and are believed to have been used as murder weapons in ancient times. In the United Kingdom these plants are not hardy and many people, therefore, keep them indoors either through the winter or all year round. I was giving an evening talk to a group and, in the break, one of them showed me a picture of herself, sitting in her sunroom, with the leaves and flowers of a large angel's trumpet draped over her shoulder. At the Alnwick Garden we put it in a cage to dramatically draw attention to its potential to bring the sound of the angel's trumpet to your ears.

The caged khat

The cage nearest the gates of the Poison Garden housed a plant that was there because of a sense of irony. Catha edulis is known

WHY ARE SOME PLANTS IN CAGES?

by a great many common names. The most common ones are all pronounced 'cat' though they may be written 'kat', 'khat', 'chat', 'qat' or 'quat'. The principal components of the plant are cathine and cathinone, psychoactive alkaloids with effects similar to amphetamine. Its effects are said to include euphoria, increased alertness and excitement, ability to concentrate, confidence, friendliness, contentment and a flow of ideas. As we shall see, however, the extent of these effects is still subject to debate.

There are a number of versions of khat's earliest discovery. According to Klaus Trenary (1997), the plant was sacred to the ancient Egyptians but no other sources suggest its use that far west. He calls it:

> A 'divine food' like royal jelly to bees, capable of 'releasing humanity's nascent divinity'. The Egyptians did not ingest khat merely to 'get high', they used it to 'trigger and impel the metamorphic process leading to a theurgic transmutation of human nature into apotheosis'. Other sources suggest it allows the lowly mortal being to be 'vergottet', or made God-like.

The medicinal uses of khat were first described in the 13th century by the Islamic physician Naguib ed-Din. He used it to treat depression. One version of khat's discovery is that it was found by a goatherd named Awzulkernayien, who observed that when the goats ate the leaves they were very alert, and he decided to try them himself. He experienced wakefulness and added awareness. There is also a legend that tells of two devout people who often spent the entire night in prayer, but frequently found themselves dropping off to sleep. They prayed to God to give them something to keep them awake. An angel appeared and showed them the khat plant, which enabled them to pursue their devotions uninterrupted.

Khat is primarily used in the southern Middle East and East Africa. Countries like Yemen, Ethiopia and Somalia are thought

IS THAT CAT DEAD?

of as the home of khat but it has, in the last 20 years or so, spread as far south as Kenya as well as west into Saudi Arabia. Emigration, especially of Somalis, has also spread khat to western countries.

The growing and sale of khat in Ethiopia offers an object lesson in how markets operate when not distorted by governments. Khat is legal in Ethiopia but not encouraged. This means growers receive no assistance in terms of subsidies or cheap fertiliser. Khat is, therefore, only grown because it can be sold at a price that is profitable to the grower.

The key features of khat are the softness of the leaves, the taste, the freshness and the power. There are numerous varieties offering different combinations of these features. Varieties with leaves staying fresh longest are more likely to be chosen for the export trade while softer chewing leaves are favoured by the higher classes even though this, generally, means a reduction in the stimulus achieved.

Because khat is legal it can be taxed and the revenue derived from khat exceeds government expenditure on healthcare, an indication of its importance. Most of the tax is collected at checkpoints on the main roads into cities where traders bringing khat to the city markets are stopped by customs officers. The Ethiopian government acknowledges that corruption and smuggling reduces the tax take but it tries to minimise this with spot checks and by moving customs men around so they do not form bonds with regular traders.

The attraction of khat as a cash crop is enhanced by its tolerance of poor growing conditions and its resistance to disease and insect attack. Khat can be grown in a large range of climatic conditions. In parts of East Africa, khat has been responsible for the development of the transport infrastructure because it needs to reach its markets quickly. Writers on khat have gone as far as to suggest that, if vaccines and medical supplies could be distributed as efficiently as khat, aid programmes would be much more successful.

The khat market, increasingly, resembles the market for any

other consumer product. In Kampala, one farmer sells khat, which he has harvested himself, from a small plantation of khat trees, and packed by hand in new banana leaves. Each package bears his signature as proof of origin and, as a result, this brand of khat, called kasuja, retails for double the price of any other in the market.

Khat is not, however, without its problems. The idea of khat as a traditional, cultural activity, which has been undertaken for hundreds of years without harm, does not stand up to scrutiny since it has spread rapidly in the last 20 or so years to parts of Africa where it was not used before. A growing concern with khat in Africa, shared by many older khat chewers, is the tendency of young people to use alcohol or cannabis after khat to alleviate the period of depression, which follows 'khating' for some people.

In the UK Somali community khat is not chewed by young people because it is thought to be the preserve of unsophisticated old men and not something new immigrants growing up in a new country should embrace. In 2006, it was announced in the House of Lords that the government would accept the recommendation of the Advisory Council on the Misuse of Drugs not to make khat itself a controlled substance since it is used by only one or two ethnic groups, mainly the Somali community, and it was felt it was unlikely to spread.

So, khat in the UK is completely legal and having it in a cage was only to draw attention to it.

Strychnos nux-vomica

Unfortunately, the other cage never had a plant in it. It was supposed to be occupied by Strychnos nux-vomica, discussed in Chapter 6, but the company who were contracted to supply the plants for the garden failed to locate many of the plants and my efforts to grow it from seed were only partially successful. It is a very large seed, about the size of a 10p coin, and takes a very long time to germinate. You have to be extremely lucky not to either let the seed dry out and die or become too wet and rot during the three months and more it takes to germinate. Eventually, on the third attempt and from a total of about 100 seeds, two germinated and began to grow. Being a pretty poor gardener, I didn't realise how quickly it would grow once it had germinated and both plants put down tap roots, which reached the bottom of their pots and left the ends of the roots exposed to the air causing the plants to die.

I don't know why the Poison Garden designers thought Strychnos nux-vomica should be kept in a cage. As there never was a plant during my time as Poison Garden Warden I have no way of knowing if it would have produced seeds in the climate in northeast England.

15

Why No Fungi?

The original plan for the Poison Garden at the Alnwick Garden was to include a 'grotto' set into the end wall, which would be used to grow mushrooms, bringing the whole issue of poisonous fungi into the tours. Budget constraints, and cost overruns, however, meant the grotto had to be dropped from the final version.

This did not, however, mean that there was no fungus in the Poison Garden. In addition to the various unintended fungi, which would display fruiting bodies in the autumn, there was one plant known to host a particular fungus, though I have to say I never did see any evidence of it.

Ergot, ergotism and some unsolved mysteries

Lolium temulentum, darnel, is a grass closely related to rye grass. For a long time it was believed to be the only variety of grass to be poisonous but it is generally accepted, these days, that it is the susceptibility of this species to invasion by the Claviceps purpurea fungus that gives it the poisonous properties. Claviceps purpurea is better known by its common name ergot, though strictly speaking the name ergot should only be applied to one stage in the development of the fungus, a substance with a long history of use and misuse and credited with having caused many thousands of deaths in its time.

Ergot fungus contains a number of harmful substances collectively called the ergot alkaloids. A number of these alkaloids have been isolated and their effects studied but this chapter is concerned with the effect of the fungus, as a whole, rather than these individual alkaloids. As well as infecting Lolium temulentum, the ergot fungus infects rye and, especially in damp years though to a lesser extent, wheat.

Outbreaks of ergot poisoning occurred frequently in medieval times, especially in a damp year when the fungus could thrive, and are said to have caused thousands of deaths. The most common effect was the constriction of the blood vessels producing a burning sensation in the limbs, known as St Anthony's Fire. This name was applied because many victims recovered after a pilgrimage to St Anthony's shrine. It is now known that removing a victim from the source of the fungus leads the condition to abate. Without this knowledge and without the money to undertake a pilgrimage, many victims developed gangrene resulting in the need for amputation. It was these amputations that formed a large part of the surgery performed at Soutra Aisle, a medieval hospital about 20 miles south of Edinburgh, where a mixture of hemlock, henbane and opium poppy was used as a sedative to keep patients asleep for up to 96 hours after the trauma of the operation.

The second well-known effect of ergot is to produce uterine contractions. It was used in 18th-century Europe by midwives to speed childbirth but it was also used to produce terminations and, in an outbreak of ergot poisoning resulting from contaminated cereal, it would produce miscarriages.

The third ergot alkaloid that would manifest itself during an outbreak of poisoning is closely related to lysergic acid diethylamide (LSD) and was the substance that led to its discovery. Dr Albert Hofmann was studying the ergot fungus to see if he could extract useful medicines from it when he synthesised LSD. He believed it could be used to treat a number of mental health problems but, when it became the recreational hallucinogenic of the hippy era, he called it 'My Problem Child' (Hofmann, 1980). Interestingly, by 2006, use of LSD had declined to such an extent that it was not possible to compile meaningful figures for its pricing across 20 UK cities covered by a survey of 'street' drug prices.

Because the proportion of the three principal alkaloids plus the many others can vary from sample to sample it is extremely difficult to be definitive about the effects of ergot poisoning so it can be attributed to a wide range of circumstances.

The true causes of the outbreak of poisoning in the French commune of Pont-Saint-Esprit in 1951 have never been fully explained but many people believe it to have been due to the ergot fungus. After a lengthy investigation of this possibility, the official enquiry concluded that the poisoning was due to a mercury-based seed fungicide. This in spite of a lack of many of the symptoms associated with mercury poisoning and the absence of a credible explanation for how the wheat believed to have caused the poisoning could have become contaminated with the mercury compound.

It is often said that Albert Hofmann, the acknowledged expert on LSD, had concluded that the Pont-Saint-Esprit outbreak was caused by mercury but he makes only one reference to it in his most important work on LSD and that is brief and in parentheses:

> [The mass poisoning in the southern French city of Pont-St Esprit in the year 1951, which many writers have attributed to ergot-containing bread, actually had nothing to do with ergotism. It rather involved poisoning by an organic mercury compound that was utilized for disinfecting seed.]
> (Hofmann, 1980)

It is also, somewhat, at odds with the account given by John G. Fuller (1968):

> When Hofmann and Stoll [Dr Stoll, a fellow researcher] heard of the poisonings in Pont-Saint-Esprit, they were further alarmed because the psychogenic symptoms were identical to those of LSD–25 ... The two men lost no time in getting

in touch with Professor Giraud at the University of Montpelier, and within days a meeting was held with many of the doctors involved in the Pont-Saint-Esprit case.

Fuller gives details of the discussion that took place with Hofmann outlining the symptoms of LSD–25 use and says:

The doctors at the meeting agreed that mercury poisoning was not evident in any manner, especially, because of the persistent lack of kidney or liver damage.

It is impossible to say whether something changed Albert Hofmann's view of the cause of the outbreak, which led him to support the mercury conclusion, but, it should be remembered that at this time LSD was hoped to become a useful treatment for mental health problems and had not yet acquired its status as Hofmann's 'problem child', so a finding that it had caused the poisoning at Pont-Saint-Esprit would be unwelcome.

In 1997 the Museo Civico di Rovereto, in Italy, published a paper that suggested that the poisoning could have been due to Aspergillus fumigatus, a mould that contains some of the ergot alkaloids. This accords with the reports that the infected wheat came from the last scrapings of a silo, which may have been the accumulation of several years.

Several hundred people were affected, with typical symptoms being insomnia, lasting several weeks, obsessive behaviour, with one victim counting his window panes for days on end and another writing for days on end, and a desire to fly. Seven people died.

A court case seeking compensation for the victims lasted several years and resulted in some compensation being paid to one of the sufferers whose case had been chosen as a test. The court, however, ruled that others who thought they might be due compensation as a result of the judgement would have to pay for their own medical examinations, which would have to show that they had been harmed. After so many years and with the perceived bias of the authorities against the villagers none of the other victims submitted themselves for testing.

Ergot fungus has been used to try and explain strange historic events. It has been suggested that the initial accusations of witchcraft made in Salem, Massachusetts, which became the subject of Arthur Miller's play *The Crucible*, were made by women suffering hallucinations after ingesting ergot. Mass hysteria then produced the full traumatic events.

Ergot has been offered as one possible explanation for the mystery of the *Mary Celeste*, the ship that, in 1872, was found drifting off Gibraltar with no sign of any of her crew, no sign of violence or illness and meals ready in the galley. In this version of the story, the ship was supplied with contaminated grain for the galley and the entire crew suffered ergot poisoning. Driven mad both by the ergot and the burning pain in their extremities they decided to jump into the sea to quell the flames of St Anthony's Fire.

Lolium temulentum, darnel, is believed by some to be the tares referred to in the 'parable of the wheat and the tares' in the Gospel of St Matthew 13: 24–30. The ancient Greeks believed that plants and animals were similar in their scientific make-up. To an extent, they thought plants were animals that couldn't move their feet. They knew, without understanding why or how, that some animals changed from one to another: caterpillars became butterflies; tadpoles became frogs, so they thought plants could do the same. Some believed that wheat could become barley. Theophrastus did not subscribe to that view but he did believe that wheat could turn into darnel. We now know that if an area of ground is very wet, wheat seeds will rot and leave the ground free for Lolium temulentum to thrive. This may be the basis for believing Lolium temulentum to be tares.

Ergot was used as one of the components of the flying mixtures used by witches, discussed in Chapter 9. These mixtures of hallucinogenic substances would be made into an ointment to be rubbed into the body, particularly of initiates, to produce the sensation of flying and this belief in the ability to fly is often recorded as a symptom of ergot poisoning incidents. It may, also, have been used as part of hallucinogenic wines made by the Greeks.

There is also the bizarre story, said to be recorded in the 'official medical records', whatever *they* are, of a woman suffering from the gangrene that ergot poisoning can cause. She was riding a horse to hospital to have a leg amputated when the horse brushed against a bush by the side of the road causing the leg to fall off.

St Anthony's Fire, the name given to the burning pain in the extremities, which is one symptom of ergot poisoning, is also the common name given to a skin infection called erysipelas. According to one visitor to the Poison Garden this can cause confusion even amongst GPs. She had a severe burning sensation in her face, which, she said, felt as though she were holding it against a lit gas fire. Her GP said it was caused by ergot poisoning resulting from eating organic rye bread, which had not been tested for the presence of the Claviceps purpurea fungus. He said there was nothing to be done except avoid all bread and he expected there to be more cases of St Anthony's Fire and other old diseases as a result of the growth of 'natural' foods.

Ergotism does not affect the face so it seems the visitor had erysipelas, which is caused by a bacterial infection and is treated with antibiotics. The visitor said that she was not given any antibiotics as the doctor said they would have no effect.

Before leaving the subject of Lolium temulentum and the ergot fungus, it is worth mentioning that John Gerard, following the recommendation of Dioscorides, says that darnel is a useful remedy for those 'that pisse in bed'.

WHY NO FUNGI?

Free food or deadly delicacies

For most people, though, the term 'fungus' means mushrooms and toadstools. 'Toadstool' is an almost entirely British idea with most foreign languages having no translation for it though Dutch has it as 'poisonous mushroom'. It is a very imprecise term because though most people would say that all toadstools are poisonous mushrooms, not all poisonous mushrooms are referred to as 'toadstools'. I prefer to stick to the term 'mushroom' or the botanically correct 'fruiting body'. Mushrooms are purely the fruit of the unseen fungus under the ground. The underside of the cap contains seed spores, which are how the fungus reproduces.

The British seem to have a phobia about mushrooms. Culturally, the belief is that the overwhelming majority of mushrooms are deadly poisonous. There are around 14,000 species of mushroom. Some are edible, even raw, and are tasty and nutritious, some are poisonous but the overwhelming majority are simply not very pleasant to eat, either because of taste or texture. It is said that only around 100 species of mushroom are poisonous but, of those, less than 20 are lethal. Even amongst the poisonous mushrooms preparation and cooking may render them edible.

The Amanita muscaria, which we'll meet later in this chapter, is said to be edible if you slice it thinly and boil it in three changes of salted water for five minutes each time before frying. I have to say I haven't read anything that suggests that the taste of the resulting meal is so exquisite as to make all that effort worthwhile but, I suppose, it's enough for some people that they are eating something most other people would avoid.

Where other cultures embrace the free dining that mushrooms found on autumn walks can provide, the British take a very precautionary attitude towards them. There are even books written giving the fullest possible detail on how to identify different types of mushroom, which end by suggesting that the safest way to collect mushrooms for eating is to go with someone who knows what they are looking for. I read these books for information about the effects of those mushrooms that are poisonous but, if I were interested in identifying edible mushrooms, I think I would be pretty upset to find that, having paid for a book and spent time reading it, its key message was go out with an expert.

Magic mushrooms

The fungus we hoped to have in the Poison Garden was Psilocybe semilanceata (the 'magic mushrooms' of 1960s and 1970s hippy fame). In 1976, reports of the plant growing in abundance in Oregon led to large numbers of young people invading pasture land to hunt for the mushroom. Its resemblance to highly poisonous types of fungi led to a number of hospital admissions. It has a long history of consumption by the human race for its psychoactive effects. It was depicted in cave paintings from Stone Age France, Mayan and pre-Mayan Central America and used as an hallucinogen in ancient Greece. In Germany, a woman who talked a lot was described as having 'eaten mushrooms'. The shape of the cap is believed to give it the power to make fairies visible.

The plant is harvested in late summer and autumn and Halloween was the occasion to consume the crop and open the door to the 'Other World' especially the Land of the Sídhe (the Land of Youth). After an apparently brief visit, the traveller returns to find hundreds of years have passed in this world.

The problem with having Psilocybe semilanceata in the Poison Garden was that mushrooms are simply the fruit of the fungus. It is theoretically possible to grow a new fungus from the spores on the mushroom but the easiest way is to dig down around the mushrooms and remove a section of fungus. All fungi can spread over quite large distances so it was felt to be unwise to have the fungus in the beds and, without the grotto, there was nowhere to keep the fungus contained.

Fly agaric

I have to say, I wasn't too disappointed that we didn't have it. The idea of having it was to add to the substance abuse message for young people but, to me, there is another fungus that offers much greater opportunities to talk about the issues involved in getting 'high'. Amanita muscaria, the fly agaric, has an unusual red cap with white spots, which, when mature, curls up at the edges to form a bowl, which was filled with milk to attract, and kill, flies.

The colourful appearance of the cap makes fly agaric a favourite in children's stories, playrooms and nurseries. The active ingredients are ibotenic acid and muscimol, which are strongly psychoactive and can cause very rapid heartbeat and a drying in the mouth. Large amounts can produce fatal convulsions. In particular, they affect the part of the brain dealing with fear. Use of the mushroom to get 'high' can lead users to place themselves in danger because of their perceived invincibility and it is said that the Viking 'Berserks', their most feared warriors, used Amanita muscaria to achieve their fearless battle hardiness.

Ibotenic acid is unlike most other substances. In almost every case, when a substance is ingested it causes a chemical reaction in the body by combining with another substance, which is naturally present. This results in the production of a third substance, which is excreted as a waste product. This process is called metabolisation. Normally, all of the ingested substance will be metabolised.

The unusual thing about ibotenic acid is that a large proportion of any ingested is excreted in the urine. The urine of someone who has eaten fly agaric mushrooms becomes psychoactive itself within about an hour of ingestion.

For the Koryak people of the Kamchatka Peninsula in the far north of the Pacific Ocean, fly agaric was the only mind-altering substance available, and then not in great quantities. The Koryak discovered that very little ibotenic acid is metabolised in producing its hallucinogenic intoxication with the vast majority being very quickly excreted from the body via the kidneys. Thus, if they were short of fresh mushrooms and wanted to get high a second time all that was required was to, as it were, pour themselves a glass. It would seem that if a friend appeared unexpectedly he would also be offered a glass as a gesture of friendship. It is also said that reindeer would eat the mushrooms so collecting and drinking reindeer pee was another way of 'enjoying' its properties.

In rituals, the order of rank of the tribe was reinforced by the ingestion of fresh mushrooms by the headman followed by progressive drinking of urine down through the social structure. It is not known if the urine retains its effects through repeated 'recycling' in this way but the junior members of the tribe would almost certainly have exhibited similar behaviour to avoid giving offence to someone from a higher level.

Fly agaric is said to be what enables reindeer to fly on Christmas Eve and it would be nice to believe that Father Christmas wears red and white because of the colour of the mushroom. Sadly, Santa's outfit has much less romantic origins; in the 1930s an advertising executive for Coca-Cola changed the traditional green to red to match the product colour.

The human race has a long history of finding and consuming psychoactive substances and there are those who argue that this search for a different view of the world induced by ingested, or injected, substances is what makes the human race successful because it is an essential part of discovery, and discovery is what moves the human race forward. That's as may be but you have to ask, if human beings are willing to drink other people's urine just to get high, is there any chance of ever eradicating the problems caused today by substances like cocaine, heroin, alcohol and tobacco?

And you have to answer, Only if we are willing to say we've progressed. Just as we no longer believe that the devil will appear if we try and pick the deadly nightshade berry so we should no longer believe that getting high is an inevitable part of human experience.

16

Is That Cat Dead?

There seems to be something built into the human mind, which makes us ask stupid questions. If ever you meet a friend in the doctor's waiting-room, I guarantee you will ask them 'How are you?' when the answer is, quite obviously 'I'm not well, and that's why I'm in a doctor's waiting-room'. And how many times, in a bar or café, have you asked, 'Is this seat taken?' when, quite clearly it isn't and you should have asked, 'Is this seat available?' Well, during the summer of 2005, the first year of the Alnwick Garden Poison Garden, one of the most popular questions asked by visitors was, 'Is that cat dead?' the answer to which is certainly not 'Yes, we thought our visitors would be pleased to see a dead cat lying in the garden'.

In their defence, however, it should be said that the cat in question, Digger, did give a pretty good impersonation, or should that be 'imcatation', of the absence of life. Her presence, and the question it provoked was, however, useful because it brought up the topic of the different effects substances, which we call poisons because they do us harm, have on different species.

In Chapter 12, I gave my opinion that it is the wrong approach to try and remove all poisonous plants from the garden in case children have contact with them. Since children will come across poisonous plants in many places other than the garden it is better to teach them, from an early age, to avoid having anything to do with substances that might be harmful. But, what happens with bees? You can't teach bees not to go to poisonous plants. Why don't they get poisoned?

IS THAT CAT DEAD?

Tell them about the honey

There's a two part answer to that question, the first part being that bees are not human so the substances that disrupt the proper function of our bodies will not, necessarily, have the same effect on bees. In Chapter 2, I explained Paracelsus' view that there are no poisons, simply substances that have an adverse reaction with a substance vital to the life processes of a particular creature. In many cases, bees can happily gather the nectar from a poisonous plant and be completely unaffected by it.

With other plants, the bees may be affected by the poisonous component but they simply do not have contact with enough of it. Bees will forage over an area with a radius of up to 5 km from their hive, further if drought or some other condition has reduced the number of flowers available. In a normal area, they will visit hundreds of different species of plant and, therefore, the total concentration of poison nectar and pollen they transport is very small. There can, however, be abnormal areas or unusual times. Plants in the Rhododendron genus are poisonous and their nectar and pollen are poisonous to bees. In the north-west of Scotland, rhododendron are particularly prolific and, early in the year, may be the only flowering plant within the range of the hive. Beekeepers in the area know that, sometimes, bees may die from visiting the rhododendron exclusively and take action to prevent the bees foraging until other plants come into flower and the danger passes.

So, bees can happily collect poison nectar and it does them no harm. But what about the honey they produce? Can that poison humans? Hopefully, you won't need me to answer that question because, I hope, if you've reached this far, you will realise that there are no heaps of corpses who have died from eating poisonous honey and the fact that we still eat honey in huge amounts is proof enough that there is no danger.

Or, rather, almost no danger.

For a long time, it has been known that the pyrrolizidine

alkaloids found in Senecio jacobaea, ragwort, can be present in honey produced by bees foraging in areas where the plant is found in high density. The assumption has always been that the concentration of these alkaloids is generally too low for the ingested dose to have any noticeable effects but, on those occasions where the concentration is high, taste becomes a factor. Honey produced with high levels of ragwort alkaloids is bitter to the taste and, in any event, has an unusually dark colour so may be identified and discarded without ever being tasted. The desire to check if this assumption was valid, however, led the Ministry of Agriculture, Fisheries and Food (MAFF) to conduct some research, the results of which were published in 1995.

This survey involved MAFF placing bee hives close to where ragwort was growing abundantly and sampling the honey formed at intervals throughout the season. In addition:

> Eight farmgate samples were obtained from small local producers to coincide with sampling from study hives. It was not possible to establish the sites of hives of the farmgate producers, the time at which these samples were bottled and whether the samples had been blended. In addition, two samples of a dark, waxy honey were donated by a small independent producer. This waxy honey was unpalatable and not suitable for blending with other honey.
> (MAFF © Crown copyright, 1995)
> (Crown copyright material is reproduced with the permission of The Controller of HMSO and the Queen's Printer for Scotland)

Of 23 samples of honey, in total, eight were found to contain ragwort pollen, including the two dark, waxy samples, which were unpalatable. The pyrrolizidine alkaloid (PA) content of these samples was assessed and then, based on estimates for normal portions of honey likely to be consumed on a daily basis, the intake of PAs for adults and children was determined using the highest concentration found in the six edible samples. The likely effect of these intakes was assessed based on work done on Symphytum spp., comfrey:

> When considering comfrey, the Committee on the Toxicity of Chemicals in Food, Consumer Products and the Environment (COT) concluded that there was sufficient evidence linking the intake of comfrey to toxic effects in humans to warrant recommending that people should not consume preparations of comfrey which contain high levels of PAs. Comfrey tablets and capsules have been voluntarily withdrawn from sale by the health food trade but comfrey leaves for use in tea infusions are still available. Tea made from comfrey leaves contains low levels of PAs. Small amounts of PAs from this source are unlikely to do any harm. For the adult consumer, estimated daily intake of total PAs from the most contaminated, edible honey (0.006 mg/person/day) was one tenth of that from one cup of comfrey leaf tea infusion (0.06 mg/cup) and is no cause for concern. Estimated total PA intakes from the honey for schoolchildren and infants were less than for adults.
>
> (MAFF © Crown copyright, 1995)
> (Crown copyright material is reproduced with the permission of The Controller of HMSO and the Queen's Printer for Scotland)

There is one area where poison honey occurs from time to time and has caused illness for humans. A plant called Coriaria arborea, tutu, is found in the Coromandel Peninsula, Eastern Bay of Plenty and Marlborough Sounds areas of New Zealand. In a particular set of circumstances it can lead to poisonous honey. There is no danger from bees visiting the flowers to gather nectar or pollen, but a vine hopper insect sucks the sap of the plant and excretes a sweet 'honeydew' containing the poisonous alkaloids of the plant. In times of drought, bees may gather this honeydew rather than plant nectar and, if the vine hoppers are abundant, they may gather enough to produce toxic honey. As well as reducing the available forage, drought also means that there has been no rain to wash the honeydew off the plant, removing it from the bees.

The vine hoppers are only active in the first four months of the year and professional beekeepers know to monitor tutu plants in their area and not sell the honey produced when the vine hopper infestation is at a high level. These days, instances of human poisoning from toxic honey are usually traced to hobby beekeepers who are either not aware of the danger or insufficiently skilled to properly manage the risk. The most recent outbreak, in early 2008, was traced, in March, to one such hobbyist after four people became unwell. On this occasion, they all survived but toxic tutu honey is said to have caused fatalities before the ways to prevent its production were understood. There have been no instances from commercially produced honey since 1974.

A thoroughbred mystery

Another incident, in which one creature was thought to have interposed itself between a poison plant and another creature and, thus, led to poisoning, occurred in Kentucky in 2001. Kentucky is the USA's breeding ground for thoroughbred horses with around 10,000 pedigree foals being produced every year. Except 2001, when over 500 foals died and approximately 2,000 other mares miscarried. At the time, the wild cherry trees, which grow in abundance in the area, were thought to be to blame. Though most often the common name for Prunus avium, wild cherry, is applied to various species of the Prunus genus, including Prunus laurocerasus, the cherry laurel. Leaves of all the Prunus genus contain cyanolipids to a greater or lesser extent and hydrolysis of cyanolipids releases cyanide and benzaldehyde. The latter has the characteristic almond smell associated with cyanide. Crushing or maceration can cause the hydrolysis to occur and Prunus laurocerasus has enough of the cyanolipids in the leaves to be used by entomologists as a way of killing insect specimens without physical damage. They seal the live insects in a vessel containing the crushed leaves.

In spite of this use – suggesting that insects are poisoned by cyanide – the culprit in the Kentucky incident was, for some time, thought to be Eastern tent caterpillars, which infested the state in large numbers that spring. The theory was that the caterpillars ate the leaves and, processed the cyanolipids into some form of cyanide and then either excreted it onto grazing or caused pollution of drinking troughs by decomposing after falling in and drowning themselves. This explanation was widely reported in both the American press and equine magazines worldwide though all the reports did say that this was a theory and further work was being done. The winter of 2000/1 was unusually hard in Kentucky and this was credited with making the wild cherry trees more toxic than normal though no evidence for this assertion was presented. This additional toxicity was used to explain why the problem had not arisen in previous years.

The problem of testing the theory that the caterpillars were the route of the poisoning was that they had turned into cocoons by the time this possibility was considered. A small study, involving 15 mares in three groups, was conducted in 2002 and showed that four of the five mares fed Eastern tent caterpillars suffered miscarriage but none of the mares in the other two groups lost their foals. The study did not attempt to define the mechanism of the abortions. The mares were fed 'starved' caterpillars, which suggests that cyanide in the faeces of caterpillars who had eaten Prunus leaves was not the cause.

At the same time that one university was putting forward the cyanide theory another suggested that hemlock might be the cause, this based on evidence of hemlock being browsed in areas where horses were kept. Again, the unusual winter was proposed as the reason why this year, and not any others, had seen problems arising.

IS THAT CAT DEAD?

Oenanthe crocata, hemlock water dropwort, is said to have the peculiar property of being more poisonous in cold weather but, from the description given and its location, it seems that it was Conium maculatum, poison hemlock, that had been browsed, and Conium maculatum needs warmth to produce the main alkaloids.

Then, in 2006, the University of Oregon published work that seems to show that vesivirus, a member of the Caliciviridae viral family, known to produce abortion in swine, marine mammals and cats and believed to have the potential to do so in many other species, was present in samples taken from dead foals and aborted foetuses from the 2001 outbreak. The virus was also found in samples of the Eastern tent caterpillar. So, it seems plant poison may not have been a factor after all.

Is that hemlock or hemlock?

Mention of hemlock brings us to one of the best examples of the difference between animal species when it comes to whether a plant is harmful or not. There are four types of hemlock. Conium maculatum is 'poison hemlock', Oenanthe crocata is 'hemlock water dropwort', Tsuga canadensis is the hemlock tree and Cicuta virosa is 'water hemlock'. Or, rather, Cicuta virosa is the species of the genus Cicuta, which is most usually called 'water hemlock' because other species in the genus such as Cicuta maculata, Cicuta occidentalis and Cicuta vagans are also known as 'water hemlock'. It is interesting that these different genera all share the common name 'hemlock' particularly as 'hemlock' is difficult to define.

Geoffrey Grigson in his *The Englishman's Flora* says that there is no definition of 'hemlock' and the word does not appear in any language other than English. 'Hemlock' is said to come from the Old English 'hymlice' or 'hymlic' and Stephen Pollington, in *Leechcraft*, suggests that this may be connected to 'hymele', the

Old English for 'hop plant'. The 'lic' or 'lice' ending in Old English is often translated as 'like'. But there is no obvious connection between the Conium and the Humulus in appearance or natural habit. 'Hym' in Old English is simply 'him' but it seems unlikely that 'hymlic' was simply 'him like'.

I don't know how the Tsuga canandensis came by the name 'hemlock' but, for the other three it is undoubtedly based on them sharing many physical characteristics, which, at the time, made people think they were related plants. In fact, their poisonous effects are very different.

The water hemlocks

Plants in the Cicuta genus contain cicutoxin, a polyacetylene alcohol, which produces nausea and vomiting with severe abdominal pain followed by death from respiratory failure or cardiac arrest, which is preceded by convulsions. The root is, generally, considered to be the most harmful part and is at its most poisonous through the winter, but the spring growth of stems and leaves has been found to be fatal to animals. Cicutoxin is highly unstable and it is the fresh plant material that is most dangerous; the one report of death following the consumption of a soup made from the plant cannot be fully verified as Cicuta and not Oenanthe crocata.

It is sometimes said that plants of the Cicuta genus are the most harmful in the USA because cattle and sheep readily graze on them and the physical similarity to parsnip and other edible plants has led to a limited number of human poisonings when raw root has been ingested in error. The problem for animals was so great that, in 1920, the University of Nevada Agricultural Experiment Station published a 'bulletin' giving details of extensive tests on the toxicity of Cicuta occidentalis. It concluded that the foliage of the plant was not sufficiently toxic to be harmful to animals in the summer or autumn but the early growth was almost as toxic as the roots and, as the early growth appears before new grass, this is the practical cause of most of the poisonings. The study showed that a particularly cold and frosty winter could lead to the rootstock becoming exposed so that the earliest growth was visible and accessible to animals. It does not conclude that this is the reason why the plant is alleged to be more poisonous in winter but that seems possible. The bulletin recommends that farmers keep animals away from the ditch banks, where the plant grows, during the early spring.

Oenanthe crocata shares the reputation of being most poisonous in the cold and its most toxic component, oenanthotoxin, a long chain polyacetylenic alcohol, is similar to cicutoxin in many ways but it is often reported to be so quick acting that death can arise very soon after the first onset of symptoms. William Rhind (1857) quotes a story that comes from the *Transactions of the Philosophical Society*.

Mr William Watson FRS, an apothecary, had written to the journal in 1744 with details of a case where two Dutch soldiers billeted at Waltham Abbey had died after eating water hemlock. It would appear this brought information on other cases and occasioned Watson to contact a Mr George Howell about a case in Pembroke. Mr Howell's reply began another article by Watson, published in 1746, about the differences between Oenanthe crocata and Cicuta virosa.

From Mr George Howell, surgeon, at Havefordwest to Mr W. Watson, Apothecary, June 1746

I have made the best inquiry I was capable of, concerning the melancholy accident at Pembroke.

Eleven French prisoners had the liberty of walking in and about the town of Pembroke; Three of them, being in the fields a little before noon, found and dug up a large quantity of a plant with its roots (which they took to be wild celery) to eat with their bread and butter for dinner. After washing it, while yet in the fields, they all three eat, or rather tasted, of the roots.

IS THAT CAT DEAD?

> As they were entering the town, without any previous notice of sickness at the stomach or disorder in the head, one of them was seized with convulsions. The other two ran home, and sent a surgeon to him. The surgeon endeavour'd first to bleed and then vomit him; but those endeavours were fruitless and he died presently.
>
> (Watson, 1746)

Remarkably, the two survivors gave no thought to what might have caused their comrade's death and served the other eight in the group with the rest of the roots for dinner.

> A few minutes after, the remaining two, who gather'd the plants, were seized in the same manner as the first; of which one died: The other was bled and a vomit with great difficulty forced down, on account of his jaws being, as it were, locked together. This operated, and he recover'd; but was some time much affected with a dizziness in his head, though not sick, or in the least disorder'd in his stomach. The other eight, being bled and vomited immediately, were soon well.
>
> (Watson, 1746)

Watson says that the symptoms described are like those reported, in 1698, in Clonmel, Ireland, when eight lads ate large quantities of a root they took to be water parsnip.

> About 4 or 5 hours later, going home, the eldest, almost of man's stature, without the least previous disorder or complaint, fell down backwards, and died convulsed. Four more died in the same manner before morning; not one of them having spoken a word from the moment the venomous particles had attacked the genus nervosum.

Watson then goes on to compare these symptoms with other cases where vomiting was reported and concludes that there is confusion between the plants, which today we know as Cicuta virosa, water hemlock, and Oenanthe crocata, hemlock water dropwort.

Watson is particularly critical of a number of the extant illustrations of the hemlocks and, for that reason commissioned Georg Dionysius Ehret to provide a drawing to accompany his article so that as many people as possible should be able to recognise Oenanthe crocata and avoid its deadly effects. In a PS to his article he notes:

> I am informed by Mr Ehret, that, in drawing the Oenanthe, which he has executed with his usual elegance and accuracy, he was obliged to have a quantity of it placed before him upon a table; when, the room being small, the effluvia therof caused him an universal uneasiness, with a vertigo, so that he was constrained to have it removed, and never after place before him but a small piece at a time.
>
> (Watson, 1746)

The publication, *Poisonous Plants in Britain and their Effects on Animals and Man* (MAFF, 1984), says that 'death is rapid and few symptoms may be seen before it occurs'. It also notes a significant difference from the Cicuta genus, which is that the toxins are much less volatile and can cause poisoning after drying and storage or when used in cooking. There seems to be no consensus on whether Oenanthe crocata causes vomiting with some case reports giving this as a symptom and others mentioning only convulsions prior to death. Of course, if deaths have occurred without symptoms, it may be that they never were attributed to Oenanthe crocata so it is difficult to rank the plant in terms of the number of deaths it may have caused over history.

I began this book with an extract from the introduction I would give at the start of each tour of the Poison Garden '... so, remember, while you're in the Poison Garden, don't touch, don't pick, don't eat and don't smell. There are plants in this garden that will make you ill just from the smell.' Oenanthe crocata is one of the plants I had in mind.

Poison hemlock

Smell is also a factor with the third of the poisonous hemlocks, Conium maculatum. Poison hemlock gives off a mouse-like odour, which is most unpleasant but does not seem to produce any physical symptoms for the majority of people. The smell does, however, deter animals from ingestion of the foliage.

Conium maculatum has completely different poisonous components from Cicuta and Oenanthe. There are five alkaloids present; coniine is the most important by far. The others are known as conhydrine, pseudoconhydrine, methyl-coniine and ethyl-piperidine. Its effects are also very different as can be seen in the most famous case of hemlock poisoning, the death of Socrates. Coniine attacks the peripheral nervous system so that the initial symptom is of numbing in the extremities, which spreads throughout the body before causing death by either paralysis of the lungs or constriction of the throat. It does not appear to have any effect on the brain and victims are said to be fully aware as the creeping paralysis overtakes them. It is, also, said that nausea and vomiting occur but there are differences of opinion as to how serious this symptom is and some case reports make no mention of it. The time to death is also subject to debate with some sources insisting death may take twelve hours or more whilst others talk in terms of one or two. The case reports that are available talk of death within anything from 90 minutes to 36 hours of ingestion.

It is generally agreed that it was poison hemlock that killed Socrates though you do see accounts, even today, claiming it was water hemlock, Cicuta virosa. The symptoms described in the *Phaedo* are those of poison hemlock – that is the slow numbing, first of the limbs but then the rest of the body with full consciousness being retained throughout. It must be said that Phaedo is a fictional character, created by Plato to describe the death of Socrates, so the account may include literary licence.

Even if the agent of his death is accepted, the circumstances surrounding it are unknown. Some sources insist on referring to Socrates' suicide although it is clear that he was under sentence of death. What is less clear is whether poison was the normal means of execution or whether, as a gentleman, Socrates was given the chance to avoid public execution by taking poison and whether he had a choice of which poison to take. Given that the quick acting Aconitum was known to the ancient Greeks, the choice of a slow killer like poison hemlock seems odd unless Socrates, as a philosopher, wanted to know what it felt like to be aware of his own impending death.

An inquest, in October 2006, heard how a biochemist used poison hemlock to commit suicide in June 2005. Wayne Calderwood collected hemlock from around his allotment and crushed it up in alcohol. Post-mortem examination showed that his throat had constricted by 70 per cent and death resulted from suffocation. At his inquest, evidence was given that he had talked about the death of Socrates. The coroner noted that it was 'a well thought out suicide and nothing further could have been done to prevent it. It is a very, very tragic case.'

But the most interesting feature of Conium maculatum, and the reason it features in this chapter, is that birds are not poisoned by it. Birds are not affected by coniine and neither metabolise it nor excrete it. This means that coniine is retained in the flesh of a bird that eats it. In parts of southern Europe a popular pastime is to net or shoot wild birds as they migrate south for the winter. Between 1972 and 1990 there were 17 cases in Italy of hemlock poisoning after eating wild birds. The toxin remained active, even in some of the cases after the meat had been frozen for storage. One person died from respiratory failure after 36 hours and another three died of kidney failure. (At this point, if I told this story during a Poison Garden tour, I would always hear some, if not all, of the visitors mutter 'serves them right'.)

IS THAT CAT DEAD?

So, is that cat dead?

So, not all animals are affected by plants in the same way, which brings us back to Digger. Digger was a feral cat, that is to say she was nobody's pet but just appeared in the Alnwick Garden at about the same time as excavations started for the Grand Cascade, hence, 'Digger'. Now, feral cats are usually scruffy and dirty and extremely unfriendly. Digger was black and white and, if she felt like it, would let you stroke her and make a fuss of her. So she was no ordinary feral cat. The other thing that made her different was that she was probably the only feral cat, ever, to have an account at the local vet.

Plants in the genus Nepeta are called 'catnip' or 'catmint' because cats seem to have a particular liking for them. It is also sometimes known as 'cannabis for cats' because the effect of Nepeta on cats is said to mirror the effect of cannabis on humans. I'm not entirely sure that's true because the initial effect on cats seems to be stimulating, making it rather more like cocaine for cats than cannabis. Shortly after exposure, however, cats do become docile and sleepy.

Its effect on humans is rather different. It is said that a human eating Nepeta will become very aggressive and angry. It has been suggested that 19th-century hangmen would eat it before work. The hangman would sit down to breakfast with the family, bounce the new baby on his knee, then eat some catmint and go and kill someone he'd never met before.

During the first summer of the Poison Garden, Digger would wander in, every afternoon, roll around in the Nepeta, tearing leaves off and scratching it up so that the plant was always in a sorry state, and then go and crawl under the patch of Artemisia absinthium, just behind the Nepeta, and have a good, long sleep, completely oblivious to the large groups of visitors standing within a few feet of her, many of them asking, 'Is that cat dead?'

17

What is Rosemary Doing in a Poison Garden?

I've already mentioned that the initial planting of the Poison Garden contained many plants that really struggled to be called poisons. Some plants were included because they were used as antidotes, but several others, like Koelreuteria paniculata, the golden rain tree, had less apparent reason to be there. When I first became involved with the Poison Garden my brief was to find stories related to the poisonous effects of all the plants.

'Anything can be a poison'

Obviously, in the case of plants like Strychnos nux-vomica or the Digitalis genus, the problem was more to do with selecting which of the many stories to use and which to leave out but with some of the others it was quite a struggle to find even one way in which the plant could be considered poisonous. Take something like Foeniculum vulgare, fennel, the herb which everyone knows has a great variety of culinary uses. With this plant, the reason given for including it was because Shakespeare refers to it as a symbol of deceit. But the straw which I clutched tightly was that fennel bears a resemblance to poison hemlock so it is possible that someone finding a plant in the wild might confuse the two. Where that possibility breaks down, of course, is that fennel has a smell like aniseed whereas poison hemlock smells like mice even without disturbing either plant.

IS THAT CAT DEAD?

While doing my research I did come across one suggestion that hemlock and fennel could cross leading to a plant that looked and smelt like fennel but contained the toxins of poison hemlock. There are actually some grounds for believing confusion could occur. It is said that Prometheus concealed the fire of the sun in a hollow plant stalk and brought it from heaven to give to the human race. In some versions of this piece of Greek mythology, the plant is said to be hemlock, in others it is fennel.

Then there was Angelica archangelica, which most people know from the small pieces of crystallized plant stem used to decorate cakes. This was included as a 'sovereign remedy for poisons and protection against contagion'. Given that most people's exposure to this plant would be limited to a few tiny pieces on a slice of birthday or Christmas cake I always felt an idiot when telling people that regular ingestion of large amounts can cause the skin to become sensitive to light.

If visitors were familiar with a plant, and even consumed it regularly, making the case for its inclusion in a poison garden became very tricky. Countless visitors spoke of eating the petals of Calendula officinalis, the pot marigold, in salads and demanded to know what the threat was. When the plants displayed their full colour in the summer, it was, usually, impossible to get the group passed the marigolds without someone insisting on finding out why they were there. The answer I gave, which is that Calendula species give off tiny amounts of pyrethrum, a natural insecticide, which is why they are recommended as a companion plant to some vegetables usually silenced the visitors but, very clearly, did not satisfy them.

Obviously, if you are working in the tourism business dealing with paying visitors, the last thing you should do is argue with them but another of the plants that didn't stay in the garden beyond the end of 2005 got me into, let's call it, a robust discussion. Rubus fruticosus is the blackberry and is most certainly not poisonous. Many people, however, remember being told, as children, not to pick blackberries after Michaelmas Day, which used to be 11th October but is now 29th September. They don't always remember what reason they were given for leaving the berries alone on the bramble bush after this date. It is said that the devil was expelled from heaven on Michaelmas Day. He was already in a pretty bad mood when he fell to earth and landed in a bramble bush, which did nothing to improve his temper. So cross was he that he spat on the blackberries and every year on Michaelmas Day the devil's spit reappears. In fact, a creature called the flesh fly lands on the fruit from about mid-October and lays its eggs in a spittle-like substance, which makes it unpleasant to eat and can cause a stomach upset.

I actually quite like that story because, like foxgloves being the home of the fairies and deadly nightshade being owned by the devil, it is an example of the ways our ancestors tried to protect their children from harm. But, one Sunday afternoon, one of my colleagues called me over to speak to a visitor who was appalled to see Rubus fruticosus bearing the Poison Garden trademark skull and crossbones label. My colleague had told the story of the devil and his spit but the visitor was not satisfied. He said he worked at Edinburgh Botanical Garden and thought it was shameful to label a plant as poisonous when it was not. Nothing I could say about how we don't always know what the mechanism of harm is for many plants or anything else could satisfy him so I had to fall back on suggesting he write to the head gardener to make his complaint.

So, when it was agreed, in the autumn of 2005, that the Poison Garden needed to be a garden of poisonous plants, I was very happy to lose almost all of the plants that had stretched our credibility. I say almost all because it was decided that it would not be sensible to remove those trees that did not merit their place and, also, to retain one plant that would surprise people because it gave the opportunity to make a very important point about substance abuse. We'll come to that plant later in this chapter.

Tall trees, tall stories

It was possible to dismiss the trees in the Poison Garden as purely there for aesthetic reasons to give a third dimension to the garden but that seemed to me a bit of a cop-out so I developed reasonable explanations for their presence. Even with the wretched Koelreuteria paniculata it was possible to say that, like a lot of fruits, the pips of the tree's fruit contained small amounts of cyanolipids, the base from which cyanide can be produced.

Of the other trees, I have already discussed the laburnum with its probably overstated reputation as a danger to children. Just across the garden from the laburnum was a Salix alba, white willow. Willow bark is a well-known painkiller and its use goes back around 2,500 years. Hippocrates is said to have prescribed willow leaf tea in 400 BC to reduce the pain of childbirth. In 1763, an English clergyman, Edward Stone, gave dried willow bark to people as a relief from rheumatic fever but it was not until 1823, when Italian scientists extracted the active ingredient and gave it the name salicin that a race started to produce this painkiller in a marketable form. The problem is that it produces stomach irritation and it was 1899 before Aspirin, a patent medicine, was launched. Aspirin is a modified form of salicin, which gives reduced gut irritation and bleeding but can still cause problems for people with ulcers or other stomach complaints.

The willow, at least, is blessed with a mass of folklore associations. Witches use willow to treat rheumatism and fever, and the old word for witches, 'wicca', may be the origin of the term wicker, applied to baskets woven from willow twigs. Wearing a sprig of willow in your hat signified rejection by a loved one. The willow tree is associated with gods and goddesses, including: Proserpina, Orpheus, Hecate, Circe, Belenus, Artemis and Mercury. One of the main properties of the willow is fertility, and due to its slender branches and narrow leaves it also became associated with the serpent. In Athens it was an ancient custom of the priests of Asclepius to place willow branches in the beds of infertile women, to draw the mystical serpents from the underworld and cure them, the connection being the phallic symbolism of the snake form itself. However, in later times this was turned around, and the willow became protective of snakes by driving them away. The ancient Spartan fertility rites of the goddess Artemis also demonstrate the willow's connection with fertility and fecundity. Here male celebrants were tied to the tree's trunk with willow thongs and flogged until the ground was fertilised with their blood and other bodily fluids.

The sound of the wind through the willow provides inspiration to poets, and Orpheus received his gifts of eloquence and communication from the willow by carrying its branches with him while journeying through the underworld. It is also associated with grief and death. The Greek sorceress Circe is said to have had a riverside cemetery planted with willow where male corpses were wrapped in ox hides and left exposed in the tops of the trees. Willow branches were placed in the coffins of the departed, and young saplings planted on their graves. The ancient Celts believed that the spirit of the dead would rise up into the sapling planted above, which would grow and retain the essence of the departed one. Watching willow grow through life eases the passage of your soul at death.

The willow's connection with water links it directly with the moon goddess. Romanian Gypsies celebrate the festival of Green George, which takes place on 23 April. A man wearing a wicker frame made from the willow represents the character of Green George. It is then covered in greenery and vegetation from the land. Pregnant women assemble around a willow tree, and each places an article of clothing beneath it. If a leaf falls onto the garment an easy delivery will be granted by the willow goddess. For the main festival, Green George uses willow branches dipped in a river to shake water on to farm animals to give good fertility in the following season.

The Euonymus europaeus is the spindle tree, given this name by William Turner who could not find an English name for it so translated the Dutch 'spilboome' given to it from its use to make spindles. Ingestion of the berries is said to produce vomiting and diarrhoea though only after an interval of up to twelve hours. John Gerard says that it has a 'loathsome smell' and a very bitter taste so ingestion is unlikely and, in any event, it is not a well-known tree because it offers a perfect winter nesting site for black bean aphids and is, therefore, unwelcome in most gardens.

The Malus 'John Downie' was discussed in Chapter 2, which just leaves two more problem trees: Hippophae rhamnoides, the sea buckthorn, and Mespilus germanica, the medlar. The information on Hippophae rhamnoides is often confusing. Some sources state that it is a strong purgative though this looks to be the result of confusion with the Rhamnus cathartica, common buckthorn, which is widely recognised as toxic. Others say it is emetic though this may be the result of the ripe berries being so acidic as to cause gagging and vomiting.

In *Native Trees and Shrubs* (Duchess of Hamilton and Humphries, 2005) the berries are described as edible but the rest of the plant is said to be poisonous. The same book, however, goes on to explain that the name Hippophae is a combination of 'hippo' meaning 'horse' and 'phae' meaning 'shine' because horses eat the foliage to improve their coats. It is Mrs Grieve who cites Henslow as saying 'in some parts of Europe the berries are considered poisonous' but this is erroneous as there is clear evidence that the berries, which remain on the plant through the winter, were used as a good source of vitamin C but, due to their bitter taste, they needed to be mixed with honey to be palatable. About the only thing you could say about this tree to justify its place in the Poison Garden was that it was evidence that nature was not to be trusted. Just as the slight sweetness of the berries of Atropa belladonna goes against the presumption that poison berries are always unpleasant to eat so the berries of the sea buckthorn go against the idea that, if a berry tastes bad, it must have a bad effect.

The other tree in the garden, Mespilus germanica, was another indicator of how nature doesn't follow the rules some people believe it should. The fruit of the medlar is inedible when ripe but, leaving it to partially rot, a process known as bletting, makes it sweet and juicy and, in this form it is used to make a meat sauce. So, it was just about possible to justify its presence but only at a very long stretch. Luckily, the tree has a story that makes visitors forget their scepticism about its credentials. The fruit has an unusual shape. It is round, about the size of a plum, but with a substantial indentation at the bottom end. Particularly if there were children around the age of eight upwards in the group I would delight in saying that, in medieval times it had another common name rather than medlar: it was known as the open arse fruit.

Why is rosemary in a poison garden?

But in addition to these non-poisonous trees, which could not be removed from the garden, there was one non-poisonous plant that was deliberately retained. Rosmarinus officinalis, rosemary, is of course known as the herb used to flavour roast lamb or to pep up roast potatoes. It is also the symbol of remembrance referred to by Shakespeare in *Hamlet* when Ophelia, in her madness, says, 'There's rosemary, that's for remembrance; pray, love, remember.' Much of the folklore surrounding the plant is associated with remembrance. It was thrown into graves as a sign that the dead person would not be forgotten, and was used to restore a failing memory. This connection with memory has been looked at scientifically in a study conducted by the University of Northumbria which looked into the effects of aromatherapy on cognitive function. One hundred and forty-four participants were exposed to the aroma of lavender or rosemary or no aroma and asked to carry out a series of tests widely used for assessing the cognitive performance of Alzheimer's patients. The study found that lavender caused reduced performance of working memory, and 'impaired reaction times for both memory and attention based tasks' whereas rosemary produced 'a significant enhancement of performance for overall quality of memory and secondary memory factors, but also produced an impairment of speed of memory'.

Rosemary is also part of the annual commemoration of ANZAC Day held on 25 April in Australia and New Zealand to remember those who lost their lives in both the Great War and the Second World War. On ANZAC Day, the wearing of small sprigs of rosemary in the coat lapel, pinned to the breast or held in place by medals is synonymous with remembrance and commemoration.

The pale blue colour of the flowers leads to another piece of folklore, which is a useful indication of how the human race has always tried to find order and meaning in the natural world. It is said that, while fleeing Herod's troops, the Virgin Mary draped her cloak over a bush of white-blooming rosemary. When she removed her cloak from the bush, the blooms had taken on the blue colour from her garment and the plant has remained blue ever since. This association with Christ is used to explain two, alleged, features of the plant. It will never grow above six feet tall, Christ's height, and it will never live for more than 33 years, Christ's time on earth. Visitors would sometimes ask me if it were true that the plant would not live longer than 33 years and I would always tell them about the visitor who called me a liar. As we were leaving the garden at the end of a tour where I had talked about the lifespan of rosemary, a visitor said to me, 'You're a liar' and before I could complain she continued, 'Before I got married at the age of 35 my name was Rosemary Bush'.

Inclusion of the rosemary in the Poison Garden was based on one feature of the plant; the aromatherapy oil made from rosemary is believed to be capable of causing a miscarriage. I say 'believed to be' because aromatherapists are trained to avoid using this and many other substances on pregnant women so there have never been any reported incidents of miscarriage. It does seem possible, however, because John Gerard cites its use in some places to 'provoke the desired sickness' (that is, bring on menstruation or, possibly, to cause a miscarriage).

And that is why the Rosmarinus officinalis was planted close to the Cannabis sativa. Because the key point about all plant poisoning and the harm caused by substance abuse is that it is not the plant, it is the way we, the human race, choose to use it. As we've seen, Papaver somniferum can be used to provide morphine and assist people to deal with terrible pain resulting from injury or disease or it can be used to produce heroin, which can dominate the life of the user and, in about 200,000 instances a year take that life. Cannabis sativa undoubtedly has beneficial effects for sufferers of multiple sclerosis and other conditions but it can dominate the life of the heavily dependent user and may increase the chance of mental illness. Nicotiana plants provide beautiful summer flowers with an attractive aroma, which is at its strongest in the evenings as the temperature drops, but when we choose to convert its leaves into tobacco it kills 5 million people a year.

When taking groups of young people around the Poison Garden, I would always stress that there is only one person in the world who can stop an individual from coming to harm because of poison plants and the substances they produce and that is the individual. If they choose to binge drink or snort cocaine, nothing anybody else says will stop them.

In taking this rather ramshackle journey through the stories of the plants in the Alnwick Garden Poison Garden plus other poisonous plants and fungi, I've tried to show where those stories about the plants may have originated and how even the oddest of them may have some small basis in fact. I'll finish with the one story that every visitor agreed was true and, as I started with a quote, I'll finish with the wording I most often used to conclude the tours: 'And my final story, and it's the story that always gets you talking amongst yourselves and forgetting all the other questions you wanted to ask is this: If there is rosemary growing in the garden, it is clear, absolute, irrefutable proof that the woman rules the household.'

References

Acosta, Christoval (1855), *Trattato di Acosta Africano medico, & chirugo*
Akre, C., Berchtold, A., Jeannin, A., Michaud, A. and Suris, J.C. (2007), Characteristics of Cannabis Users Who Have Never Smoked Tobacco, *Archives of Pediatrics & Adolescent Medicine*
Allsopp, S., Dillon, P. and McLaren, J., Swift, W. (2008), Cannabis potency and contamination: a review of the literature, *Addiction, The Journal of the Society for the Study of Addiction*
Anderson, M., Del Castillo, J. and Rubottom, G.M. (Jan 1975), Marijuana, absinthe and the central nervous system, *Nature*
Arnold, Wilfred N. (1992), *Vincent Van Gogh: Chemicals, Crises and Creativity*

Barnes, T., Burke, M., Jones, B., Lingford-Hughes, A., Moore, T. and Zammit, S. (2007), Cannabis use and risk of psychotic or affective mental health outcomes: a systematic review, *The Lancet*
Beverley, Robert (1705), *History and Present State of Virginia*
Booth, Martin (2005), *Cannabis, A History*
Breaux, T.A., Kubalia, T., Lachenmeier, D.W., Nathan-Maister, D., Schoeberi, K. and Sohnius, E.M. (2008), Chemical composition of a vintage absinthe with special reference to thujone, fenchone, pinocamphone, methanol, copper and antimony concentrations, *Journal of Agricultural and Food Chemistry*

Camden, W. (1806), *Britannia*

Casida, J.E., Ikeda, T., Narahashi, T. and Sirisoma, N.S. (2000), a-Thujone: y-Aminobutyric acid type A (GABA-A) receptor modulation and metabolic detoxification, *Proceedings of the National Academy of Sciences*
Comitas, L. and Rubin, V. (1975), *Ganja in Jamaica*
Commerford Martin, T. and Dyer, F.L. (1910), *Edison, His Life and Inventions*

de Graaf, R., Monshouwer, K., van Dorsselaer, S. and van Laar, M. (2007), Does cannabis use predict the first incidence of mood and anxiety disorders in the adult population? *Addiction, The Journal of the Society for the Study of Addiction*
DEFRA (2004), *Code of Practice on How to Prevent the Spread of Ragwort*, Farm Focus Division
Duchess of Hamilton, J. and Humphries, C. (2005), *Native Trees and Shrubs*
Duplais, P. (1855), *A Treatise on the Manufacture and Distillation of Alcoholic Liquors*

Felter, H.W. (1922), *The Eclectic Materia Medica*
Forrester, R.M. (1979), Have you eaten laburnum? *The Lancet*
Fuchs, L. (1542), *De Historia stirpium*
Fuller, John G. (1968), *The Day of St Anthony's Fire*

Gerard, J. (1598), *Great Herbal or General History of Plants*, 1633 edition (revised by Johnson, T.)
Grieve, M., *A Modern Herbal* (1931)
Grigson, G. (1958), *The Englishman's Flora*

Hickman, M., Jones, P., Kirkbride, J., Macleod, J. and Vickerman, P. (2007), Cannabis and schizophrenia: model projections of the impact of the rise in cannabis use on historical and future trends in schizophrenia in England and Wales, *Addiction, The Journal of the Society for the Study of Addiction*
Hofmann, Albert (1980), Trans. Ott, J., *LSD - My Problem Child*

REFERENCES

La Guardia Committee Report (1944), *The Marihuana problem in the city of New York*, Mayor's Committee on Marihuana, New York Academy of Medicine, City of New York

Maccinis, P. (2004), *Poisons – From Hemlock to Botox and the Killer Bean of Calabar*

MAFF UK (1984), *Poisonous Plants in Britain and the Effects on Animals and Man*

MAFF UK (Feb 1995), *Surveillance for pyrrolizidine alkaloids in honey*, Information Sheet No. 52

Max, B. (1992), This and that – an artefactual alkaloid and its peptide analogs, *Trends in Pharmacological Sciences*, 13(9): 342–345

Millspaugh, Charles F. (1887), *America Medicinal Plants*

Montgomery, Marshall (July 1920), *The Modern Language Review*, 15(3)

Palatnick, W., Meatherall, R., Sitar, D. and Tenenbein, M. (1997), Toxicokinetics of Acute Strychnine Poisoning, *Clinical Toxicology*, Vol 35, Issue 6

Parkinson, J. (1629), *Theatrum Botanicum*

Patouillat, Trans. 'T.S.M.D' (1737), *Journal of Philosophical Transactions*, 40

Pliny (AD 77–79), Trans. Jones, W.H.S., *Natural History*

Rhind, W. (1865), *History of the Vegetable Kingdom*

Schenk, Gustav, Trans. Bullock, M. (1956), *The Book of Poisons*

Smyth, F. (1980), *Cause of Death: The Story of Forensic Science*

Stedman, Dr J. (Letter to Pringle, Dr J.) (1751), *Journal of Philosophical Transactions*, 47

Trenary, Klaus (1997), *Catha edulis*: sacred plant of the ancient Egyptians. www.lycaeum.org

Turner, William (1551), *A New Herball*

von Bingen, Hildegard, Trans. Hozeski, B.W. (2001), *Physica*

Watson, William FRS (1746), *Transactions of the Philosophical Society*, 44

Withering, William (1785), *An Account of the Foxglove*

WUV'N ACRES www.wuvie.net/castorbeans.htm